THE
TRIAL
OF
DARWIN'S GOD

RANDOMNESS:
SCIENTIFIC TRUTH OR RELIGIOUS DOCTRINE?

GEORGE COLEMAN CONRAD

ISBN: 1-4392-5556-3
ISBN-13: 9781439255568

Visit www.booksurge.com to order additional copies.

For Lily and Bridger

In hopes that they may be taught
only science in science class.

TABLE OF CONTENTS

PREFACE

What in the world is Darwin's God and why on Earth should Darwin's God be placed on trial? Two good questions.

Answer No. 1

Charles Darwin was a brilliant scientist who came to believe that the ultimate reality of the universe and life itself is chance and randomness. Darwin's God is the Accidental God.

God, by definition, is ultimate reality. There are only two possible choices for the ultimate reality of the universe and life itself: accident or intelligent design. God is either accidental or purposeful. One or the other.

Organized religions and individuals who believe in the ultimate reality of intelligent design, a Purposeful God, comprise more than 90% of America's population. On the other hand, over 90% of our most elite scientists believe in the ultimate reality of chance and randomness, the Accidental God – Darwin's God. The religion that worships Darwin's God is known as scientific atheism.

Answer No. 2

The First Amendment of the United States Constitution bluntly provides:

> "Congress shall make no law respecting an establishment of religion, or prohibiting the free exercise thereof . . . . "

This is known as the 'Establishment Clause', and the Supreme Court of the United States has rigorously enforced that prohibition. Indeed, the Supreme Court has determined that the Fourteenth Amendment of the Constitution extends the prohibition of the Establishment Clause to include the actions of each of the fifty States as well as those of the Federal government.

The Supreme Court has consistently held that the Establishment Clause prohibits any religious doctrine from being taught in public school science class because the doctrine cannot be supported by confirmable scientific evidence.

The explanation for ultimate reality that is now routinely taught in public school science class is randomness. Until now the Supreme Court has not addressed the issue of whether the **randomness explanation** for the creation and adaptive evolution of life is supported by confirmable scientific evidence.

If the randomness explanation is actually supported by confirmable scientific evidence it should be taught in science class. If the randomness explanation is not supported by confirmable scientific evidence it must be deemed to be doctrine of the religion of scientific atheism. As such, it must be expelled from science class, like the doctrines of all other religions, as a violation of the Establishment Clause.

Darwin's God needs to be placed on trial in order to determine if there is, in fact, confirmable scientific evidence to support the randomness explanation that is a core belief of the religion of scientific atheism.

* * *

Charles Darwin published his seminal work, *On the Origin of Species By Means of Natural Selection,* in 1859. That book presented confirmable scientific evidence that all living creatures, including the species *Homo sapiens,* were descended from ancient antecedent ancestors. In short, confirmable scientific evidence was presented that contradicted the creation story contained in the Bible. Darwin presented compelling evidence supporting the evolution of life.

For millennia science and religion had been in lockstep. Now, for the first time, science and religion presented sharply different versions of creation and life.

For nearly a century now there has been a swirling controversy in America concerning the teaching of evolution in the science classes of our public schools. Early in the 20th century the legislatures and public agencies of several of these United States enacted laws in different forms aimed at ensuring that Darwin's scientific theory would not be taught in public schools. Since then, our Federal Courts have examined the legality of teaching the theory of evolution in public schools and have consistently ruled that evolution is a valid scientific theory. Darwin's theory of evolution may be taught in science class, but any theory that includes a hint of traditional religious views must be barred from the classroom as a violation of the Establishment Clause.

Over the course of fifteen decades the confirmable scientific evidence for evolution by means of natural selection has kept on mounting and mounting. Yet, the controversy continues. Why?

Simply put, modern scientists have expanded Darwin's theory of evolution to now insist that science has **proven** that the creation and evolution of all life can be explained in accordance with solely natural processes that are undirected and without purpose. In short, our most brilliant scientists now proclaim that they have gazed upon the face of God and have found that ultimate reality has no purpose and provides no direction. Based on this authoritative proclamation public school science textbooks now explain that all aspects of life can be factually explained without the need for a Purposeful God. The ultimate and overarching explanation now taught in science class is **randomness**, also known as accidentalness. And, this ultimate and overarching explanation is established doctrine of the religion of scientific atheism.

Elite and mainstream scientists devoutly maintain that the only mechanism through which natural selection evolves new species, and evolves adaptive traits in all species, is random mutation of genetic DNA. Science has demonstrated quite convincingly that evolutionary change occurs as a result of genetic DNA mutation. But, have scientists provided confirmable scientific evidence in support of the explanation that these mutations are **random**? Whether or not such DNA mutations are random is central to answering the question of whether or not ultimate reality is accidental or purposeful.

Teaching in science class that nothing exists beyond the natural world is not to be taken lightly. If children are taught that the only purpose for our existence is to pass on our DNA, the take home lesson for living life is the adoption of a philosophy called **nihilism** wherein:

- There is no basis for objective morality;
- There is no intrinsic value in human life;
- There is no purpose and meaning in life beyond that which we simply make up.

So, if we are going to teach our children in science class that **random** mutations are responsible for the creation and evolution of all life, that explanation needs to be supported by a lot of confirmable scientific evidence.

God, by definition, is ultimate reality. For Charles Darwin and the vast majority of the scientific elite in America today, the foundation for ultimate reality is **randomness**. Randomness is without intention or direction or purpose. Randomness is dependent only on mistakes and accidents. In other words, ultimate reality is accidental. Most elite scientists in America today firmly believe in this Accidental God. They pay homage to Darwin's God.

* * *

The purpose of this book is to place Darwin's God on trial. If the evidence produced during this trial provides confirmable scientific evidence that Darwin's God is in fact real, then the ultimate and overarching explanation of **randomness** should be taught as true in the science classes of our public schools. If confirmable scientific evidence is lacking then Darwin's God should be treated the same as the Gods of all other religions and expelled from science class as a violation of the Establishment Clause.

I cannot imagine a more serious subject than the question of purpose and meaning in life. The meaning of life and our purpose in living prescribes our worldview. It defines how we see things in life. Randomness and purposefulness represent polar opposite worldviews. And, in the final analysis either one or the other represents ultimate truth and ultimate reality.

Each of us should base our worldview on the best evidence possible.

Yet, we need not be solemn and stern as we explore serious subjects. We should also have a little fun as we pursue truth.

This little book was written as a purely fictional account of a trial. It is an attempt to explore a very serious subject by interjecting a little good humor along the way. No offense is intended to anyone. I hope the reader finds the presentation in this manner to be educational, thought provoking, and fun.

For those readers who would like to explore this most serious subject in more depth, I invite you to read a more comprehensive book that I have written, entitled *The Accidental God Fallacy: A Primer for Perspective on Scientific Atheism.*

INTRODUCTION

Let me introduce myself. My name is Sabastian Story. Most people call me Shorty even though I have never liked that very much.

I'm just a smidgen over 6 feet 6 inches tall so most folks think that calling me Shorty is clever. That's not clever. That's sad country humor. But I'm stuck with it.

I live in a small town in western Kentucky, just north of the Tennessee State line. The name of the town is Hazel, just like the color of my eyes.

Shorty Story is quite a handle for a newspaper reporter. But, that's what I am. And, bowing to popular sentiment, my byline is Shorty Story.

I'm the science writer for the *Hazel Clarion,* a fine publication that is printed twice weekly for distribution on Wednesdays and Saturdays. I'm also the sports writer, do the weekly column on gardening tips, and am responsible for the accuracy of the obituaries. The owner and editor of the paper, my Uncle Rob, does the rest. You can pretty well figure out how I got the job.

Last summer I was assigned by Uncle Rob to cover a trial that was taking place not too far south of here in a town called Dayton, Tennessee. Uncle Rob usually does the court stuff himself. But, because this trial really involved science, the science writer got the call. And I'm glad I did.

You see, this trial represented the latest in the courtroom struggles that have been evolving for nearly a century now about the line between science and religion in public schools.

I reported the courtroom proceedings during the course of the trial and many of my readers urged me to compile the gist of those writings into a book, so that is just what I have done. I am not an accomplished author, so please bear with me if the presentation is not as refined as you are used to. I'm doing my best.

I think the best way to proceed to tell the story is to begin with a little background. You see, without some background the latest court case will just seem to be a jumble. That doesn't do a thing for really understanding the issues involved. So, to aid in understanding the issues, the first chapter will review the scientific developments leading to the Modern Synthesis of Evolutionary Theory. The second chapter will then review the legal road that led to the juncture at which the Trial of Darwin's God in Dayton began. The third chapter will get straight to the 21st century Dayton trial. Let's begin.

CHAPTER 1

The Road to the Modern
Scientific Theory of Evolution

Evolution is a term that we use almost every day. Evolution is defined simply as the process of change in a certain direction, an unfolding.

Our human methods of transmitting information have evolved over time from simple speech, to writing, to sending information over long distances, first by smoke then by electronic signals. The transmission of information through electronic signals began with the telegraph codes. It then evolved to telephone, radio, television, cellular telephones, personal computers and the Internet. Today we can instantly twitter our most personal observations to our heart's content to anyone, anywhere in the world.

To the contrary, the transmission of information within all living things is done through the work of an information bearing little molecule called DNA. The structure of DNA that transmits information in living organisms has not evolved. Yet, living organisms that form and operate in accord with the information instructions contained in the DNA molecule have evolved to include millions of different species of living creatures that now reside on our planet Earth.

Evolution is the process of change. The mechanism of change that results in the evolution of living organisms is the subject of this book.

From Natural Theology to Natural Selection

As with many things modern, the roots of evolutionary theory are found in ancient Greece. Democritus and Epicurus, members of the philosophical school of atomists, pronounced that there was a **continuity in nature and that changes were random. Those are the core components of modern evolutionary theory.** But, in the march of time Aristotelian teleology became the prevailing viewpoint of Christian Europe.

In the Greek language, *telos* (end) combines with *logos* (reason) to make *teleology* (the study of end reason or purpose). Teleology is firmly based on the idea that the universe and the things of the universe are designed for a purpose.

For Aristotle, teleology was the study of final causes. The final cause is the reason for which a natural phenomenon was created. Aristotle believed that natural things developed in order to realize the end that is internal to their own nature.

Following in the footsteps of Isaac Newton, scientists of the 17th and 18th centuries devoted themselves to finding mechanistic explanations for natural phenomena that operated within a 'Clockwork Universe'. All natural phenomena, both physical and biological, were viewed as machines designed by an intelligent agent. The intelligent agent was called God and the prevailing scientific belief was **natural theology**.

In the late 18th and early 19th centuries an English naturalist and clergyman, William Paley, published his teleological arguments for the existence of God in such writings as *A View of the Evidence of Christianity* and *Natural Theology*. Paley's writings were required reading for undergraduate students at the University of Cambridge. Charles Darwin was such a student.

Natural theology proclaimed that the nature of God can be best understood by studying the natural world that he created. Paley maintained that God's design was evident in all living things of nature. In *Natural Theology* he used a watch as an analogue for a living organism. He argued that logic compelled the conclusion that both must have been designed, and design necessarily requires an intelligent agent. As he stated in this abridged quote:

> ". . . when we come to inspect the watch, we perceive that its several parts are framed and put together for a purpose . . . the inference we think is inevitable, that the watch must have had a maker; that there must have existed at some time and at some place or other, an artificer or artificers who formed it for the purpose which we find it

actually to answer; who comprehended its construction and designed its use."

Paley contended that the only way to account for the vast variations and adaptations of living organisms on the planet, each of which is far more complicated than a watch, is to recognize that each was created by an intelligent designer.

As Paley observed the world around him he saw harmony and happiness everywhere.

"It is a happy world after all. The air, the earth, the water, teem with delighted existence. In a spring noon, or a summer evening, on whatever side I turn my eyes, myriads of happy beings crowd upon my view."

Thomas Malthus, a British economist, saw things in sharp contrast to Paley. He observed the world of the British poor and saw them living in wretched conditions. Their lot was one of human despair. Malthus pronounced that human population would grow exponentially (2,4,8,16) while the food supply would grow only arithmetically (2,3,4,5). He predicted that the growth of population would quickly outpace the food supply. Population could only be kept in balance with food supply if human population is subjected to plagues, starvation, and infanticide. The weak will die. The strong will survive. Those people in the human population who can best adapt to their change in fortune will be those who survive to raise their children. Others will die of starvation and disease. Only the fittest will survive. (Note: World population in 1900 was 1.8 billion; in 1950 it was 2.1 billion; in 2000 it was 6.0 billion; in 2050 it is projected to be 9.2 billion.)

By the 19th century most scientists had abandoned the Aristotelian view that species of living organisms are immutable, never change, and exist forever. The new sciences of geology and paleontology were revealing a 'fossil record'. That record contained evidence of past extinctions of great beasts who obviously do not roam the Earth today. And, anatomists were amassing a great deal of data that revealed amazing similarities among many species, including the species *Homo sapiens*. The evidence had become compelling that living organisms had evolved. The question became: 'What was the **mechanism** that guided the evolution of life?'

Most scientists then believed in a vitalistic evolutionary theory proposed by a Frenchman, Jean-Baptiste Lamarck. He used the example of the evolution of a giraffe's long neck. He theorized that a giraffe really strived (worked and worked) to stretch his neck to reach the desired leaves residing on a higher branch. Because of this striving the giraffe's offspring would inherit a slightly longer neck. As this striving was passed on generation to generation the final result was the greatly elongated neck of the modern giraffe. The crux of such a theory of evolution was that evolutionary change over time was directed in accordance with the requirements of the organism evidenced by its striving. In Lamarckian evolution the mechanism for change over time is a **vital inner drive.**

In the mid-19th century two Englishmen concluded that the real mechanism that accounted for the evolution of living organisms over time was not some vital inner drive. As they independently developed their theory of an evolutionary mechanism, they were each much influenced by the writings on populations by Thomas Malthus. Malthus had developed his ideas by looking at humans not as individuals but as groups of individuals and introduced the concept of population dynamics. And he had introduced the concept that would become ever-after associated with evolution - survival of the fittest. This British economist had provided both Russel Wallace and Charles Darwin with the mechanism of evolution that has become the centered heart of modern biology - **natural selection**. Charles Darwin published his book on natural selection before Wallace. Thus, Darwin today is a household name while Wallace is little known.

Darwin's Theory of Evolution by Natural Selection

In 1859 Charles Darwin, an Englishman, published his famous book supporting modern evolutionary theory. It is called *On the Origin of Species by Means of Natural Selection or the Preservation of*

Favoured Races in the Struggle for Life. Today he is widely recognized as the 'father of evolution'.

In his seminal work, Darwin began by taking great pains to describe his theory. At core **Darwin's theory of evolution** is this. Living organisms inherit traits from their parents that may or may not help them in their struggle for survival. Those traits that prove to be valuable to a species will be retained in the gene pool of the species because those individuals possessing those traits will be the most likely to reproduce and pass them on. In essence, **heritable traits that prove valuable in the struggle for the survival of the species will be retained.** Much better put, in Darwin's own words:

> "Owing to this struggle for life, any variation, however slight and from whatever cause proceeding, if it be in any degree profitable to an individual of any species, in its infinitely complex relations to other organic beings and to external nature, will tend to the preservation of that individual, and will generally be inherited by its offspring. The offspring, also, will thus have a better chance of surviving, for, of the many individuals of any species which are periodically born, but a small number can survive. I have called this principle, by which each slight variation, if useful, is preserved, by the term of Natural Selection, in order to mark its relation to man's power of selection."

By Darwin's lifetime, man had been selectively breeding plants and domestic animals for literally thousands of years. Man was the intelligent agent who 'selected' the new and improved plant or animal by selective breeding. Darwin proposed that the same type of thing was going on with non-domesticated plants and animals all the time. Yet, no intelligent agent was required. The mechanism of 'natural' selection was one of random mutations. **Darwinian evolution is descent with modification by natural selection over evolutionary time.** It is **natural selection working on random mutations** (i.e., genetic modifications due to mistakes or accidents). The modifications selected from random mutations are not selected by some vital inner-drive as suggested by Lamarck. **The selected modifications happen purely by chance and are retained in the gene pool only if they prove useful for survival of the species.**

Let me be clear about Darwin's teaching. Charles Darwin never used the term 'random mutation'. Darwin described his theory of Natural Selection as a theory of 'descent with modification':

> "Natural selection can act only by the preservation and accumulation of infinitesimally small inherited modifications, each profitable to the preserved being."

Darwin believed that candidate modifications resulted from 'accidental deviation' and that such variability arose 'from the indirect and direct action of the external conditions of life'. Charles Darwin never used the modern term 'random mutations', but his writings make it crystal clear that Darwin's adaptive change mechanism over time was rooted in **randomness**. He explained that biological change was not purposeful or intentional. He just knew that, at core, it was accidental.

With the unfolding of the science of genetics in the 20th century, evolutionary change mechanisms have been expanded to include such things as genetic drift and gene flow, as well as the natural selection of mutations. However, the only mechanism that can result in **adaptive modifications** over time is still believed to be **random mutation**. The modern mainstream evolutionary belief is succinctly stated by Harvard University Professor, Ernst Mayr, one of the founders of the modern synthesis of evolutionary theory:

> "Ultimately, all variation is, of course, due to mutation."

Darwinian evolution, and the modern mainstream evolutionary belief, are all about adaptive traits. The contention is that the evolution of adaptive traits by random mutations enhances the survivability of individuals within a species population who inherit those traits. And, on a large-enough scale, the evolution of adaptive traits by random mutations may result in the development of a new species. For Darwinian evolution and the modern mainstream evolutionary belief, **random mutation is the key**.

Post-Darwin Discoveries in Genetics and Cellular Biology

Genetics, broadly speaking, is the branch of biology that deals with heredity and variation of living organisms. More narrowly, in common parlance today, it is the study of heredity linked to genetic makeup. Cellular biology is the study of the composition of living cells.

When Charles Darwin shocked the world with his theory of evolution by natural selection he and his scientific colleagues were wholly ignorant of genetics and cellular biology. Darwin believed that 'pangenes' from each and every part of the body came together to form the egg and the semen that fertilized it. And Darwin thought that living cells were filled with a kind-of mushy protoplasm.

Genetics

Variation in living organisms had been observed since first there were observers. And, since the beginning of recorded history people had been selectively breeding domesticated animals and edible plants. A speedy champion mare was mated with a speedy champion stallion to produce a new horse race champion three years later in an ancient analogue of today's Kentucky Derby. The 'trait' of swiftness was inherited by the colt. In early agricultural efforts, farmers cross-pollinated plants with the 'traits' of disease-resistance and hardiness and thereby produced more abundant and healthier crops for the following year's harvest. It was common knowledge that 'heritable traits' were passed on to offspring through breeding.

For thousands of years scientists believed that the factors responsible for trait inheritance simply blended together as a result of sexual reproduction. But, if such blending occurred then, over time, the distinct variations that we see in species of living things would vanish. After all, if you pour and stir two different buckets of paint together, eventually all of the blended paint will be identical. That clearly does not happen in living organisms. Quite a conundrum for science. How then is variation maintained in natural species? Answers needed to be found.

Gregor Mendel was an Austrian monk who lived in the mid-19th century. He was greatly interested in solving this conundrum. He wanted to find the secret of how variation is naturally maintained in living populations. So, he started to investigate and experiment. The subjects he chose for experimentation were pea plants. And, he experimented a lot, using more than 28,000 pea plants. He made 287 crosses between 70 different purebred pea plants. While in his lifetime no one knew what a gene was, his experimental results, records and theories describing 'Mendelian traits' accord him the niche in history as the 'father of genetics'.

In carefully controlled experiments, Mendel tested the 'blending hypothesis' in peas. He crossed true-breeding purple-flowered pea plants with true-breeding white-flowered pea plants. His results disproved the blending theory. The offspring plants did not have blended whitish-purple flowers. All offspring of the 1st generation had purple flowers. In the 2nd generation some offspring had white flowers and some had purple flowers, but none had flowers of a blended color. The white flower trait that had apparently disappeared in the 1st generation had somehow reappeared in the 2nd. How could that be?

Further experimentation and documentation of generation after generation of pea plants led Mendel to develop a scientific hypothesis that accurately described the working of genes and chromosomes long before genes were ever discovered.

Cellular Biology

In the 17th century Anton van Leeuwenhoek of Holland developed the first practical microscope. Through it he observed bacteria and the circulation of blood in capillaries. A primitive microscope first revealed to Robert Hooke, an English physicist, small cubicles, or rooms, in dead bark cork. He called them cells. Cellular biology thus began at that early date. By the later part of the 19th century improvements in the microscope instrument allowed biologists to observe the nuclei within living cells. Walter Sutton and Theodor Boveri observed that the nuclei of cells split in half during the formation of gamete sex cells. By the early part of the 20th century they published their theory that each gene has a specific location on a chromosome in the cell nucleus.

As the 20th century dawned, Thomas Hunt Morgan began to investigate how mutations occurred in living organisms and **discovered that most mutations result in dysfunction**. Morgan was an American zoologist who experimented a lot with fruit flies. Through his experiments he discovered the linkage of chromosomes in cells and the phenomenon of 'crossing over' that occurs when two chromosomes of a homologous pair exchange equal segments with each other at the first part of meiosis (sex cell division). He mapped the genes on the chromosomes of the fruit fly. In short, he **discovered that genes are responsible for heritable traits and that genes are linked in a series on chromosomes.**

In the 1930's a Russian geneticist, Theodosius Dobzhansky, worked in the laboratory of Thomas Hunt Morgan. He studied the genetics of populations by examining the differences that emerged between different isolated populations of the same species. He discovered that all members of a specific species do not have identical genes. He believed that what kept a species distinct was simply sex. Members of the same species could mate and have offspring. Organisms outside the species could not. In 1937 he published *Genetics and the Origin of Species* in which he explained that mutations occur in genes all the time. And, most importantly, he observed that:

"The process of mutation is the only source of raw materials of genetic variability and hence of evolution."

By the middle of the 20th century much had been observed and learned about the nature of proteins at work in living cells. By then it was known that cell chromosomes consisted of both proteins and DNA. Most scientists believed that proteins held the key to understanding the chemical basis of genetics, and were responsible for passing on genetic information through cell reproduction. Oswald Avery then **discovered that DNA, not proteins, was the carrier of genetic information.** His discovery prodded Francis Crick and James Watson to further investigate the mysteries of DNA.

In 1953 they used X-ray diffraction to discover the structure of the DNA molecule. Their three dimensional model of DNA shows the coiling of two sides of a genetic ladder into a double helix. The sides of the ladder (backbone) repeat a phosphate-sugar bond over and over. The rungs of the ladder comprise the variable part of the molecule. The sequencing of the base pairs of nucleic acids along the rungs of the ladder provides the genetic information. They further discovered that the exact sequencing of the four bases in DNA and in RNA served to provide construction instructions for building functional proteins out of 20 discrete amino acids. In short, they broke the genetic code. Crick and Watson were awarded the Nobel Prize in Medicine in 1962.

The Modern Synthesis of Evolutionary Theory

When the brilliant life scientists today talk about evolution it seems almost like they are trying to make the matter confusing. The terms that they use are often used in different contexts and are used to describe different aspects of natural phenomena in seemingly contradictory ways. The problem lies not with these brilliant scientists. The problem results from the immense amount of knowledge that they have gained about how living things operate and change. It is really hard to explain all this stuff in a simple manner.

Darwin's theory of evolution by natural selection has been expanded to today include not only random mutations, but such things as 'genetic variation', 'random genetic drift', 'gene flow', 'speciation by reproductive isolation', and 'punctuated equilibrium'.

Darwin's idea was that all organisms that have ever lived have evolved from a common ancestor. Virtually all biologists agree with that today. Darwin believed that the mechanism for evolutionary change was natural selection through randomness. Today, biologists have provided much greater elaboration regarding the randomness mechanism. In 1986 D.J. Futuyama, a renowned biologist, published *Evolutionary Biology* which includes this synopsis of the modern synthesis of evolutionary theory:

> "The major tenets of the evolutionary synthesis, then, were that populations contain genetic variation that arises by random (i.e., not adaptively directed) mutation and recombination; that populations evolve by changes in gene frequency brought about by random genetic drift, gene flow, and especially natural selection; that most adaptive genetic variants have individually slight phenotypic effects so that phenotypic changes are gradual . . . that diversification comes about by speciation, which normally entails the gradual evolution of reproductive isolation among populations; and that these processes, continued for sufficiently long, give rise to changes of such great magnitude as to warrant the designation of higher taxonomic levels (genera, families, and so forth)."

It is most obvious that modern evolutionary biology has discovered a great deal about the evolution of living things since the days of Charles Darwin. But, most importantly, the modern mainstream evolutionary belief maintains that the evolution of all adaptive change is the result of **random** mutations of genetic DNA. These scientists believe that random mutations of genetic DNA produce all adaptive traits which result in either:
- increased survivability within a species population; or
- on a large-enough scale, in the development of a new species.

All adaptive change is, therefore, explained as the result of random mutation that occurs in one of two ways:
- a mistake during DNA replication in a sex cell; or
- an alteration of genetic DNA in response to some radioactivity or cosmic rays or some poison in the environment.

In short, the very foundation of the modern elite and mainstream evolutionary belief is **random mutation caused by an accident**.

CHAPTER 2

Legal Views on Teaching Only Science in Science Class

For most of the past century there has been a swirling controversy in America concerning the teaching of evolution in the science classes of our public schools. The legislatures and public agencies of several of these United States enacted laws in different forms aimed at ensuring that no blasphemous scientific theory be taught as true in public schools. State and Federal courts have consistently ruled that a wall of separation must be maintained between the state and any religion. It all started with an Englishman named Charles Darwin.

From the publication of Darwin's theory of evolution and throughout the first quarter of the 20th century, a societal divide was widening in America. On one side of the divide remained the traditionalist believers in the long-held biological view of speciation that conformed with the creation story of Adam and Eve portrayed in the Bible. On the other side were the modernists who held Darwinian evolution and Freudian psychology in the highest esteem. As the divide widened and deepened a new wave of Christian fundamentalism grew in strength in America's southern states. The little town of Dayton, Tennessee was destined to become the location of the first great legal clash between science and religion in the 20th century.

By 1925 bills had been introduced in the legislatures of 15 states that would ban the teaching of evolution in public schools. Tennessee became the first state to pass such a bill and have it signed into law by the Governor. In March of 1925 the State of Tennessee passed the following bill into law:

"AN ACT prohibiting the teaching of the Evolution Theory in all the Universities, Normals and all other public schools of Tennessee, which are supported in whole or in part by the public school funds of the State, and to provide penalties for the violations thereof.

Section 1. *Be it enacted by the General Assembly of the State of Tennessee,* That it shall be unlawful for any teacher in any of the Universities, Normals and all other public schools of the State which are supported in whole or in part by the public school funds of the State, to teach any theory that denies the story of the Divine Creation of man as taught in the Bible, and to teach instead that man has descended from a lower order of animals.

Section 2. *Be it further enacted,* That any teacher found guilty of the violation of this Act, shall be guilty of a misdemeanor and upon conviction, shall be fined not less than One Hundred ($100.00) Dollars nor more than Five Hundred ($500.00) Dollars for each offense."

This legislative act became known as 'The Monkey Law', and shortly after its passage legal battle erupted in the little town of Dayton, Tennessee.

State v. John Scopes (aka 'The Monkey Trial') – 1925

John Scopes taught general science at Dayton High School. He taught, among other things, Charles Darwin's theory of evolution, as portrayed in the State-approved textbook used in the classroom entitled *Hunter's Civic Biology.*

Scopes was criminally charged with violating the Tennessee statute and the famous 'Monkey Trial' shortly followed during the summer of 1925.

While both the prosecution and defense included several distinguished and able attorneys, the large press corps depicted the trial as a battle between two of the country's most prominent lawyers: William Jennings Bryan for the prosecution and Clarence Darrow for the defense. The presiding judge for the *Scopes* trial was John T. Raulston.

The trial began with a motion from the defense to quash the indictment on state and federal constitutional grounds, but Judge Raulston denied the defense motion.

Trial testimony quickly revealed the fact that Scopes had indeed used the *Hunter's Civic Biology* textbook and taught that humans and other animals had evolved from single-celled organisms.

The trial lasted for over a week and trial coverage by the press served to provide the country with a science lesson. But, in the end, no constitutional clarity was provided by this bizarre case. Darrow for the defense asked the jury to return a guilty verdict so that the case could be appealed to the Tennessee Supreme Court. They did, Judge Raulson fined Scopes $100, and Darrow appealed.

The following year the Tennessee Supreme Court reversed Scopes' conviction on the technicality that the fine should have been set by the jury, not the Judge. The Tennessee Supreme Court did not remand the case back to the trial court for further action because Scopes no longer taught or resided in Tennessee. Rather, they dismissed the case outright. With the dismissal there was no chance for constitutional clarity by the United States Supreme Court. More than forty years would pass before such clarity would be forthcoming.

Epperson v. Arkansas – 1968

Although the *Scopes* trial in Tennessee provided no constitutional clarity on teaching evolution in public school classrooms, it did serve to deter anti-evolution efforts. Of the fifteen states that had introduced anti-evolution statutes by the summer of 1925, only Mississippi and Arkansas actually enacted laws restricting the teaching of Darwin's theory of evolution. A challenge to the Arkansas statute would finally provide some constitutional clarity in the year of 1968. The Arkansas statute that had been State law since1929 read as follows:

> "*Doctrine of ascent or descent of man from lower order of animals prohibited.* – It shall be unlawful for any teacher or other instructor in any University, College, Normal, Public School, or other institution of the State, which is supported in whole or in part from public funds derived by State and local taxation to teach the theory or doctrine that mankind ascended or descended from a lower order

of animals and also it shall be unlawful for any teacher, textbook commission, or other authority exercising the power to select textbooks for above mentioned educational institutions to adopt or use in any such institution a textbook that teaches the doctrine or theory that mankind descended or ascended from a lower order of animals.

Teaching doctrine or adopting textbook mentioning doctrine – Penalties –Position to be vacated. – Any teacher or other instructor or textbook commissioner who is found guilty of violation of this act by teaching the theory or doctrine mentioned in section 1 hereof, or by using, or adopting any such textbooks in any such educational institution shall be guilty of a misdemeanor and upon conviction shall be fined not exceeding five hundred dollars; and upon conviction shall vacate the position thus held in any educational institutions of the character above mentioned or any commission of which he may be a member."

This legislative act remained dormant for nearly four decades until Susan Epperson was employed by the Little Rock school system to teach 10th grade biology at Central High School.

For the academic year 1965-1966 the Little Rock school district prescribed that high school biology teachers must use a textbook containing a chapter on Darwinian evolution. Upon assuming her teaching position Ms. Epperson was faced with a true dilemma. The school district required her to teach from a textbook including a chapter condemned by the State statute. If she did not use the prescribed textbook she faced discipline by the school district and possible termination of employment from her teaching position. If she did use the prescribed textbook she would commit a criminal offense and be subject to job termination for violating the statute. So, she sought judicial relief in Arkansas Chancery Court seeking a declaratory judgment that the Arkansas statute is void and injunctive relief to prevent the school district from dismissing her from her teaching position.

The Arkansas Chancery Court resolved her dilemma for a short while by ruling that the Arkansas statute was unconstitutional and void. The State of Arkansas appealed the ruling and the Supreme

Court of Arkansas reversed. The Arkansas Supreme Court sustained the State statute as a valid exercise of the State's power to specify the public school curriculum.

Epperson appealed that decision to the United States Supreme Court. The nation's highest court held that the Arkansas statute violates the Fourteenth Amendment, which embraces the First Amendment's prohibition of laws respecting an establishment of religion. In short, the Arkansas statute was an unconstitutional violation of the First Amendment's religious Establishment Clause.

When our nation's highest court issued its opinion and ruling in 1968 it went a long way toward clarifying the standard for teaching only science in science class. The following highlights from the *Epperson* case are illustrative.

- "Fundamentalist sectarian conviction was and is the law's reason for existence . . . . There is and can be no doubt that the First Amendment does not permit the State to require that teaching and learning must be tailored to the principles or prohibitions of any religious sect or dogma.

- By and large, public education in our Nation is committed to the control of state and local authorities . . . . On the other hand, the vigilant protection of constitutional freedoms is nowhere more vital than in the community of American schools . . . . The First Amendment does not tolerate laws that cast a pall of orthodoxy over the classroom.

- The State's undoubted right to prescribe the curriculum for its public schools does not carry with it the right to prohibit, on pain of criminal penalty, the teaching of a scientific theory or doctrine where that prohibition is based upon reasons that violate the First Amendment.

- The State may not adopt programs or practices in its public schools or colleges which 'aid or oppose' any religion. This prohibition is absolute. It forbids alike the preference of a religious doctrine or the prohibition of theory which is deemed antagonistic to a particular dogma.

- Neither a State nor the Federal Government can pass laws which aid one religion, aid all religions, or prefer one religion over another.

- Government in our democracy, state and national, must be neutral in matters of religious theory, doctrine, and practice. It may not be hostile to any religion or to the advocacy of no-religion; and it may not aid, foster, or promote one religion or religious theory against another or even against the militant opposite. The First Amendment mandates governmental neutrality between religion and religion, and between religion and nonreligion."

Edwards v. Aguillard – 1987

Almost twenty more years would pass before the United States Supreme Court provided further clarity regarding the standard for teaching only science in science class. In 1987 the United States Supreme Court issued its legal ruling on the legality of a Louisiana statute that was known as the *Balanced Treatment for Creation-Science and Evolution-Science in Public School Instruction Act* that was signed into Louisiana law in 1982 (aka 'The Creationism Act').

The Creationism Act did not prohibit the teaching of evolution in science class. Rather, it forbade the teaching of evolution unless creation science was also taught. No school was required to teach evolution or creation science. However, if one was taught the other also had to be taught.

The stated secular purpose of the Louisiana Creationism Act was to protect academic freedom by furthering the goal of 'teaching all of the evidence'.

Aguillard brought suit in Federal District Court seeking to invalidate the Act because it violated the religion Establishment Clause of the Constitution and sought summary judgment. The District Court granted summary judgment to invalidate the Act. The Court of Appeals affirmed and the United States Supreme Court agreed. The following highlights from *Edwards v. Aguillard* provide further clarity by our nation's highest court on the issue of teaching only science in science class.

- "The Court has been particularly vigilant in monitoring compliance with the Establishment Clause in elementary and secondary schools. Families entrust public schools with the education of their children, but condition their trust on the understanding that the classroom will not purposely be used to advance religious views that may conflict with the private beliefs of the student and his or her family. Students in such institutions are impressionable and their attendance is involuntary. The State exerts great authority and coercive power through mandatory attendance requirements, and because of the students' emulation of teachers as role models and the children's susceptibility to peer pressure.

- In no activity of the State is it more vital to keep out divisive forces than in its schools.
- In this case the purpose of the Creationism Act was to restructure the science curriculum to conform with a particular religious viewpoint.
- The First Amendment does not permit the State to require that teaching and learning must be tailored to the principles or prohibitions of any religious sect or dogma."

Kitzmiller v. Dover Area School District – 2005

Nearly twenty more years would pass before further clarity would come from the Federal Courts regarding teaching science in science class. This time the final decision would be made by the United States District Court for the Middle District of Pennsylvania. No appeal from the District Court's decision would be made by the School District.

In 2004 the Dover Area School District of Pennsylvania determined that, commencing in January 2005, biology teachers in the District's high school would be required to read the following disclaimer to students in biology class:

> "The Pennsylvania Academic Standards require students to learn about Darwin's Theory of Evolution and eventually to take a standardized test of which evolution is a part.
> Because Darwin's Theory is a theory, it continues to be tested as new evidence is discovered. The Theory is not a fact. Gaps in the Theory exist for which there is no

evidence. A theory is defined as a well-tested explanation that unifies a broad range of observations.

Intelligent Design is an explanation of the origin of life that differs from Darwin's view. The reference book, *Of Pandas and People*, is available for students who might be interested in gaining an understanding of what Intelligent Design actually involves.

With respect to any theory, students are encouraged to keep an open mind. The school leaves the discussion of Origins of Life to individual students and their families. As a Standards-driven district, class instruction focuses upon preparing students to achieve proficiency on Standards-based assessments."

Tammy Kitzmiller, parent of a biology student in the School District, et al, filed suit challenging the constitutional validity of the above School District teaching requirement (the Intelligent Design Policy). A bench trial was conducted before District Court Judge John Jones III. On December 20, 2005 Judge Jones issued his 139-page decision, ruling that the Dover School District disclaimer was unconstitutional. The ruling barred Intelligent Design (ID) from being taught in the District's high school science classes. The ruling in this case added no further clarity by our nation's highest court on the issue of teaching only science in science class. However, Judge Jones' lengthy decision did provide some insight on teaching Intelligent Design in public school science classes. The following highlights from Judge Jones' ruling are illustrative.

- "The Supreme Court has instructed that the word 'endorsement is not self-defining' and further elaborated that it derives its meaning from other words that the Court has found useful over the years in interpreting the Establishment Clause. The endorsement test emanates from the 'prohibition against government endorsement of religion' and it 'precludes government from conveying or attempting to convey a message that religion or a particular religious belief is *favored or preferred.*'

- The disclaimer's plain language, the legislative history, and the historical context in which the ID Policy arose, all inevitably lead to the conclusion that Defendants consciously chose to change Dover's biology curriculum to advance religion.

- The evidence at trial demonstrates that ID is nothing less than the progeny of creationism.
- Since the scientific revolution of the 16th and 17th centuries, science has been limited to the search for natural causes to explain natural phenomena. This revolution entailed the rejection of the appeal to authority, and by extension, revelation, in favor of empirical evidence. Since that time period, science has been a discipline in which testability, rather than any ecclesiastical authority or philosophical coherence, has been the measure of a scientific idea's worth.
- 'Methodological naturalism' is sometimes known as the scientific method. Methodological naturalism is a 'ground rule' of science today which requires scientists to seek explanations in the world around us based upon what we can observe, test, replicate, and verify.
- The National Academy of Sciences (hereinafter 'NAS') was recognized by experts for both parties as the 'most prestigious' scientific association in this country. NAS is in agreement that science is limited to empirical, observable and ultimately testable data: 'Science is a particular way of knowing about the world. In science, explanations are restricted to those that can be inferred from confirmable data – the results obtained through observations and experiments that can be substantiated by other scientists. Anything that can be observed or measured is amenable to scientific investigation. Explanations that cannot be based upon empirical evidence are not part of science.'
- ID is predicated on supernatural causation. Stated another way, ID posits that animals did not evolve naturally through evolutionary means but were created abruptly by a non-natural, or supernatural, designer. ID fails to meet the essential ground rules that limit science to testable, natural explanations.
- ID is reliant upon forces acting outside of the natural world, forces that we cannot see, replicate, control or test, which have produced changes in this world. While we take no position

on whether such forces exist, they are simply not testable by scientific means and therefore cannot qualify as part of the scientific process or as a scientific theory.

- A final indicator of how ID has failed to demonstrate scientific warrant is the complete absence of peer-reviewed publications supporting the theory. The evidence presented in this case demonstrates that ID is not supported by any peer-reviewed research, data or publications.

- It is our view that a reasonable, objective observer would, after reviewing both the voluminous record in this case, and our narrative, reach the inescapable conclusion that ID is an interesting theological argument, but that it is not science.

- Our conclusion today is that it is unconstitutional to teach ID as an alternative to evolution in a public school science classroom."

Modern Legal View on Teaching Only Science in Science Class: A Synopsis

We have now reviewed the significant legal decisions to date regarding teaching only science in the science classrooms of public schools. The courts have provided a pretty clear picture of what is and what is not permissible – what standards must be maintained in order to pass constitutional muster regarding the religious Establishment Clause contained in the First Amendment. In short:

- Teaching science in science class must be restricted to those explanations that can be inferred from confirmable data – the results obtained thorough observations and experiments that can be substantiated by other scientists. Anything that can be observed or measured is amenable to scientific investigation. Explanations that cannot be based upon empirical evidence are not a part of science.

- Explanations that cannot be inferred from confirmable data are beyond science. They are referred to as either metaphysical or religious. They have no place in science class.

- Explanations that are beyond science are prohibited as aiding or opposing a religion. The prohibition is absolute. It forbids alike the preference of a religious doctrine or the prohibition of theory which is deemed antagonistic to a particular dogma.

- Government in our democracy, state and national, must be neutral in matters of religious theory, doctrine, and practice. It may not be hostile to any religion or to the advocacy of no-religion; and it may not aid, foster, or promote one religion or religious theory against another or even against the militant opposite.   The First Amendment mandates governmental neutrality between religion and religion, and between religion and nonreligion.
- The First Amendment does not permit the State to require that teaching and learning must be tailored to the principles or prohibitions of any religious sect or dogma.
- Intelligent Design (ID) fails to meet the essential ground rules that limit science to testable, natural explanations.   It is, therefore, unconstitutional to teach ID as an alternative to evolution in a public school science classroom.

The elements of this synopsis have become widely known by the state legislatures and local school districts throughout America. It is also widely known that the *Kitzmiller* case cost the local school district a lot of money. Kitzmiller's legal fees had to be paid by the Dover Area School District at a cost between one and two million dollars. The battle over teaching evolution in public schools seemed to be over. But, perhaps another test case would be necessary to ensure that only science is taught in the science classes of our public schools.

The Remainder of this Book is Fictional

In 2009 the State of Tennessee enacted the *'Only Science in Science Class Act'*. The passage of that new Tennessee State law was quickly followed by the adoption of a companion policy by the Dayton, Tennessee School District entitled the *'Only Science in Science Class Policy'*. The adoption of that well-intentioned act and that well-intentioned policy would soon provide compelling evidence in support of the rule of unintended consequences.

Shortly after their enactment the second *Scopes* trial revisited the little town of Dayton. And, once again, it started on a hot summer day.

CHAPTER 3

John Scopes v. Dayton School District
The Judge's Overview

"All rise for the Honorable John T. Raulston, Presiding Judge for the United States District Court for Southern Tennessee. This Court is now in session." The surly bailiff thereby called to order the prelude to the trial that would become widely known at the outset as 'Scopes 2' and, by the conclusion, as 'The Trial of Darwin's God'.

Judge Raulston took his seat behind the raised bench and hand-motioned for all to be seated. The bailiff verbalized the Judge's gesture and a hush descended on the courtroom.

The Judge adjusted his reading glasses as he began to address the assemblage, reading from notes on the desk before him.

"The trial that will commence tomorrow is the case of *John Scopes v. Dayton School District*. The trial will be conducted as a bench trial. That means that the trier of facts in this case will not be a jury but, rather, will be the trial Judge – namely me. The attorneys for both Mr. Scopes and the School District have agreed to this bench trial proceeding. Such is not unusual in a case like this one. When the determinative facts are closely related to constitutional questions of law the interests of justice may be better served by a bench trial. Both parties in this case have agreed to that."

A long pause settled over the Judge's bench as papers on the desk were examined and arranged. He then continued.

"In the further interest of justice and of time, both the Plaintiff and the Defendant, in pre-trial conference, have agreed to many important stipulations.

Stipulations represent an agreement between the parties that certain matters relevant to the case at bar are not contested. For purposes of the trial the stipulations are considered proven facts. The stipulations are intended to narrow the issues to be decided in this case. Such stipulations are in the interest of justice, have been entered into the case record and are accepted by this court as true in making determinations that will bear on the outcome of this case."

* * *

Before proceeding with the Judge's remarks it may be useful for the reader to become familiar with the stipulations referenced by Judge Raulston. He did not read them aloud to the courtroom because, as the trier of fact, he alone had to understand and use them. The thoughtful reader may well be more than somewhat interested in what the actual stipulations in this case were. So, I have included this digression. Here are the most important stipulations.

- In science, explanations are restricted to those that can be inferred from the confirmable data. Scientific explanations are bound by that limitation.
- Evolution by natural selection of genetic mutations is an explanation that can be inferred from confirmable data. It is a valid scientific theory supported by abundant evidence derived from the fossil record, homology between species and the historical development of DNA.
- The scientific method requires that explanations of phenomena that cannot be falsified must be excluded from science.
- Science does not consider issues of 'meaning' or 'purpose' or 'ultimate explanations'.

Now, back to the courtroom.

* * *

Judge Raulston continued with his explanation of stipulations.

"During the course of this trial I expect that learned counsel may make reference to such stipulations. As that occurs I will make a point of ensuring that the referenced stipulations are accurate as reflecting the record as agreed. I will not hesitate to promptly correct counsel if they are not. When stipulations have been accepted by this Court as true, they cannot and will not be misrepresented in this Court."

Acknowledging those seated in the crowded courtroom, he continued from the notes before him.

"Members of the press are well represented in this courtroom today. You are welcome and you will, of course, freely report anything you like about these proceedings. That is the inalienable right of a free press in a free society. You are guaranteed that right by the First Amendment of our Constitution. Interested spectators not affiliated with the press are also welcome. But let me be very clear. There will be no mockery or disturbance within this courtroom. Justice demands civility and decorum. If civility and decorum are not maintained this trial will be conducted with neither press nor interested spectators present. And, the free press will necessarily be limited to the public record kept of this trial."

Judge Raulston concluded reading from his prepared notes, removed his reading glasses, and began to address the courtroom in a more cordial and less-formal fashion.

"The coincidences in this trial are overwhelming. The 'Scopes 2' and 'Scopes in Reverse' labels given to it by some news reports already are actually quite clever. But, the facts are simply coincidental—nothing else."

Sitting straighter in his chair he continued.

"My name is John Raulston. The name of the trial Judge in the 1925 'Monkey Trial' was John Raulston. The lead lawyer for the School District is William Jennings Bryan, the same name as the famous lawyer for the prosecution in the 1925 trial. The lead lawyer for Mr. Scopes is Clarence Darrow, the same name as the famous lawyer for the defense who represented Mr. Scopes in 1925. The setting for this trial is in Dayton, Tennessee. The setting for the 1925 Scopes trial was also in Dayton. The 1925 trial began on July 10th. Tomorrow is July 10th. This trial starts tomorrow. And of course, the science teacher in this case is John Scopes, the same name as the science teacher who was prosecuted by the State of Tennessee in 1925."

Shaking his head he proceeded.

"It is hard to believe that all of these things are coincidences, but that is exactly what they are. None of us namesakes are related in any manner to the 1925 cast of characters in the first Scopes trial. And, conducting this trial in Dayton is due to the simple fact that shortly after the beginning of the 21st century the Federal District Court for the Southern District of Tennessee was relocated to the same town where John Scopes lived in 1925. Coincidences. All coincidences. And, nothing can be, or in this courtroom will be permitted to be, inferred from such coincidences. Justice demands civility and decorum. Both will be maintained in this courtroom and no mockery will be abided."

Some stirring and coughing and a little subdued laughter emanated from the gallery as the Judge paused again for nearly a minute as he consulted the papers before him. He then continued.

"The trial being conducted in this courtroom is a civil lawsuit. It concerns an issue of constitutional importance.

The Plaintiff, John Scopes, is seeking what is known as declaratory relief. Mr. Scopes contends that the *Only Science in Science Class Policy*' of the Dayton School District is violative of the religious Establishment Clause of the United States Constitution.

That School District Policy requires biology teachers to teach that the natural selection of **random** genetic mutations is a scientific explanation for the creation and evolution of life on Earth. That **randomness explanation** is contained in two books written by the National Academy of Sciences (NAS) that the

School District requires Mr. Scopes to use in his science class. Mr. Scopes wants this Court to issue a judgment ruling that the School District Policy must be declared to be null and void because it is in violation of the First Amendment of the Constitution."

Reading directly from the paper before him he continued.

"Declaratory relief is granted very sparingly in this country by its Federal Courts. The circumstances presented must exhibit that substantial harm will be suffered by the Plaintiff unless such relief is granted. This Court has found that Mr. Scopes' circumstances exhibit the requisite substantial harm required as a predicate to considering such relief. That is why this trial is proceeding at this time.

The circumstances facing Mr. Scopes present a very real dilemma. As a scientist he believes that if he follows the *'Only Science in Science Class Policy'* of the Dayton School District he will be forced to violate the *'Only Science in Science Class Act'* of the State of Tennessee. If he violates the State Law he will lose his job and likely go to jail. If he violates the School District Policy he will be faced with summary dismissal for cause by the District and thereby lose his teaching position."

* * *

Before proceeding with the Judge's remarks it may be useful for the reader to examine the background and actual language of both the State Law and the School District Policy at issue in this case. So, another digression follows. Please bear with me.

In the autumn of 2008 America began to experience the up-close and personal effects of a serious economic recession that had begun about two years before. The long-inflated housing price bubble burst and the financial credit market both at home and throughout the world began to unravel. Elected politicians were faced with the need to be seen as taking action to solve the dual problems of rising unemployment and economic dislocation affecting their constituents.

Legislators of the State of Tennessee caucused to develop plans to entice clean and progressive businesses to locate within the State and thereby increase the employment and tax bases. Providing jobs for the citizens of the Volunteer State was a prime motivator.

Additional tax incentives were enacted. Land was dedicated for progressive economic development and made available at little or no charge to relocating companies. Yet, to compete effectively with other progressive states legislative leaders determined that it would be greatly beneficial to take some action to improve Tennessee's image. Many folks still considered Tennessee as a backwater part of the country. That image attached following the infamous 1925 Scopes 'Monkey Trial'. Images are hard to change. Something had to be done to show the corporate boardrooms of America that the Volunteer State was not populated by a bunch of Bible-banging Christian fundamentalists who still believe that planet Earth is only 6,000 years old and that teaching evolution should be banned from the science classes of public schools.

The enlightened leaders of the State wanted to send a message loud and clear that Tennessee is a progressive and modern State into which any progressive, modern business would want to locate and prosper and wherein school children are taught only real science in public schools.

The 2005 Pennsylvania District Court ruling in the *Kitzmiller v. Dover Area School District* case was widely known by State legislatures, including the Tennessee legislature. In that case the Federal District Judge ruled that Intelligent Design (ID) could not be taught in science class. So, the legislative leadership determined that including some of the language and conclusions reached in that Pennsylvania case into Tennessee State Law would be just the ticket to help dispel a backwater image.

Thus motivated, the Tennessee General Assembly enacted and the Governor of Tennessee signed into law the *'Only Science in Science Class Act'* that reads as follows:

'Section 1: PUBLIC SCHOOL DISTRICTS OF THE STATE are prohibited from authorizing the use of textbooks or other teaching materials that aid or oppose any religion. This prohibition is absolute. It forbids alike the preference of a religious doctrine or the prohibition of theory which is deemed antagonistic to a particular dogma.

Section 2: IT SHALL BE UNLAWFUL for any teacher in any of the public schools of the State to teach or endorse the doctrine of any religion. The teaching of science in science classes shall be limited to only natural explanations that can be inferred from confirmable data whose results can be observed, tested, replicated and verified. Scientific explanations are restricted to those that can be inferred from confirmable data. Inferences that cannot be inferred from confirmable data are deemed to be doctrine of a religion and cannot be included in the science curriculum of public schools. Explanations that cannot be based on empirical evidence are not a part of science.

Section 3: POSITION TO BE VACATED AND MISDEMEANOR CRIME – Any teacher who is found guilty of violation of this act by teaching the doctrine of any religion shall be convicted of a misdemeanor crime and punished accordingly; and upon conviction shall vacate the teaching position held.'

The Board of Education of the Dayton School District was a progressive and enlightened Board. All of its members were delighted with the new State Law and wanted to act quickly to reinforce strict compliance with the State Law by all faculty members within the District. One of the Board members pointed out that the portion of the State Law that specified the limitation for teaching science in science class was a paraphrase of the materials contained in two books published by the National Academy of Sciences (NAS). And, the Judge's ruling in the 2005 Pennsylvania case specifically noted that experts for both parties recognized the NAS as the 'most prestigious' scientific association in this country.

The Board was motivated to strictly comply with the new State Law. After lengthy discussions they determined that the best way to ensure compliance was to require biology teachers in the District's

High School to teach the explanations of evolution contained specifically in two books written by the NAS, the country's elite scientific authority. Thus motivated, the Board adopted the *'Only Science in Science Class Policy'* that reads as follows:

> 'WHEREAS, the School Board recognizes the legislative significance of the State Law recently enacted entitled *'Only Science in Science Class Act'* as furthering progressive education in our State, and
>
> WHEREAS, the Board wishes to fully hold the science teachers of the District to the letter of the law established by that Act, now therefore
>
> BE IT RESOLVED that all biology teachers employed by the District must teach evolution and use, as supplemental materials concerning the origin and evolution of life, the following books published by the National Academy of Sciences (NAS):
>
> (1) *Science and Creationism: A View from the National Academy of Sciences,* Second Edition (1999) ISBN 0-309-53224-8, and
>
> (2) *Science, Evolution, and Creationism* (2008) ISBN 0-309-10586-2.
>
> BE IT FURTHER RESOLVED that the failure to use the explanations contained in the NAS books for the origin and evolution of life shall result in summary dismissal from employment by the District.
>
> BE IT FURTHER RESOLVED that every biology teacher employed by the District shall be personally provided a copy of this policy and advised of its contents and consequences.'

John Scopes has been employed as a biology teacher at Dayton High School for the past three years. During his employment at the school he has always used the materials concerning the evolution of life contained in the standard textbook specified by the School Board. He has always referred to genetic mutations without specifying that they must be random. And he has never included anything in his classroom regarding the origin of life.

Mr. Scopes has concluded, as a scientist, that there is an abundance of scientific evidence that the evolution of life is the result of the natural selection of genetic mutations. But, he is not aware of any scientific evidence that supports the conclusion that genetic mutations are in fact **random**.

For Mr. Scopes the explanations contained in the NAS publications, which insist that the origin and evolution of life are the result of **random** genetic mutations, are not supported by 'confirmable data whose results can be observed, tested, replicated and verified' as specified by State Law.

Mr. Scopes sincerely believes that he will be in violation of the *'Only Science in Science Class Act'* if he teaches the NAS randomness doctrine contained in the two books specified in the School District's *'Only Science in Science Class Policy'*.

When John Scopes was personally provided a copy of the School District Policy and advised of its contents he fully comprehended his dilemma. He then filed this lawsuit.

End of digression.

* * *

Judge Raulston finished reading from the paper before him and removed his reading glasses. He then summed up the matter in lay language that the gallery appreciated.

"To be blunt, Mr. Scopes finds himself on the horns of a real dilemma. Without declaratory relief he will be subject to losing his job one way or the other and he might end up going to jail for doing something he does not want to do but is forced to do. Declaratory relief is the proper remedy to resolve this dilemma.

The matter boils down to this: Are the explanations contained in the two NAS books which maintain that the origin and evolution of life are the result of the natural selection of **random** mutations scientific explanations? Stated another way: are those explanations supported by confirmable data whose results can be observed, tested, replicated and verified as required by State Law? Are those explanations scientific or religious in nature? If they are scientific Mr. Scopes' dilemma is resolved and he can follow the School District Policy with no fear of violating State Law. If they are religious they are barred from being taught in public school as a violation of the First Amendment Establishment Clause and the School District Policy is null and void."

Judge Raulston gazed upon the assembly without speaking for what seemed a full minute. He made eye contact with many of the observers throughout the courtroom. He then concluded.

"The single focus of this trial will be to factually determine if the **randomness** explanations contained in the two NAS books that the School District requires Mr. Scopes to use in his High School biology class are indeed scientific. Those books explain that the origin and evolution of life are the result of the natural selection of **random** mutations. So, the single focus of this trial will be to factually determine if that explanation of **randomness** is scientific or religious.

In pre-trial conference both the Plaintiff and the Defendant have agreed that, following opening arguments, each day of the trial will be devoted to the examination of limited scientific subject matter. The purpose of that is to allow the trier of fact, namely me, to have sufficient time each day to not only hear the testimony but to study the matter thoroughly from my notes and the transcript.

I've taken the opportunity today to explain to the press and interested observers the scope and focus of the upcoming trial. It is my hope that today's overview will serve to maintain civility and decorum in this courtroom. Both will be maintained in this courtroom, and no mockery will be tolerated.

With that, Court is adjourned until 9:00 am tomorrow morning when the trial of *John Scopes v. Dayton School District* will begin with opening arguments from the Plaintiff and the Defendant. I bid you good day."

CHAPTER 4

The Trial – The First Day
Opening Arguments and the Scientific Method

The first day of the 'Scopes 2' trial began with the same opening ritual performed by the bailiff. Judge Raulston handled some housekeeping matters before calling on the lead attorney for the Plaintiff.

"Mr. Darrow, if you are ready to proceed, it is now time for your opening statement."

Clarence Darrow rose from his seat at the Plaintiff's table and approached a lectern before the Judge's bench. He took notes with him. Darrow, short in stature but large in presence, was 55 years of age, a seasoned and well-respected trial lawyer. He dressed in a conservative gray flannel suit complete with vest and pocket-handkerchief. He was accompanied in his presentation by bifocal glasses and a cane. He had suffered a serious hip injury some years before that left him with a limp and need of assistance in walking. He arranged his notes on the lectern and spoke.

Plaintiff's Opening Argument

"Thank you your honor. My client, Mr. Scopes, did not choose to make a big deal about teaching science in science class. Mr. Scopes is a scientist through and through. He has a master's degree in biology from Vanderbilt University. And, he has chosen to spend his working life not in the pursuit of money but, rather, in teaching the path and truth of science to youngsters in public school.

Before his tenure of the past three years with the Dayton School District Mr. Scopes taught in Nashville public schools for fifteen years. He has always in his teaching career been dedicated to teaching the truth of science and has pointedly avoided stretching the boundaries of science beyond what can be supported by scientific evidence. He is ethically bounded in his teaching by the fundamental principle that only science should be taught in science class. That is what he sincerely believes and that is what he intends to do in his high school biology classroom. However, with the enactment of the *'Only Science in Science Class Policy'* the Dayton School District is now demanding that he include materials in his biology class that clearly stretch the boundaries of science far beyond what can be supported by scientific evidence."

Clarence Darrow turned from the lectern to acknowledge the presence of a distinguished member of the National Academy of Sciences seated at the Defense table and continued.

"The National Academy of Sciences is the most prestigious scientific association in America today. That has been true for a long time. The NAS was founded at the time in American history when we were engaged in a great civil war. And the NAS is still the preeminent body to officially advise the United States Congress on all things scientific. The NAS membership represents the scientific elite of this country. When they provide scientific explanations they speak with great authority and we trust that what they teach us is based on solid scientific evidence."

He continued after sorting through some of the materials before him.

"The fact that the overwhelming majority of the NAS membership are **atheists** detracts not one iota from their scientific credentials. However, it may well be an explanation for why the NAS publications contain materials that go far beyond the boundaries of science and enter the realm of the metaphysical and religious.

Clarity of understanding requires a clear definition of terms. So, let me clearly define two essential terms that I will be using throughout this trial concerning the realm of the metaphysical and religious:

- Metaphysical simply means anything beyond the physical. Any explanation that cannot be supported by confirmable scientific evidence is, by definition, metaphysical.

- Religious simply means belief in a particular ultimate reality, also known as God.

A 1998 survey of the NAS membership revealed that 93% of the scientists who responded to the survey did not believe in a Purposeful God. Those scientists believe in the God of chance and accident and randomness – Charles Darwin's God.

God, by definition, is ultimate reality. Charles Darwin believed in the ultimate reality of chance and accident as the fundamental explanation for the creation and adaptive evolution of life on Earth. Darwin's God was the Accidental God. And, the elite membership of the NAS today overwhelmingly believes in the Accidental God – Darwin's God.

The belief of brilliant scientists that ultimate reality is accidental is not founded on the discoveries of science. That belief is solidly founded on the 'idea' that nothing beyond nature can possibly exist. The belief is called **scientific atheism** and it has become a fervent religious belief. It is different and more insidious than all other religious beliefs for it is cloaked in the 'authority of science'.

The National Academy of Sciences first provided a bright light to illuminate their belief through the publication *Science and Creationism: A View from the National Academy of Sciences*. The 2nd edition was published in 1999. Then, in 2008 the National Academy of Sciences published a book entitled *Science, Evolution and Creationism*. These publications provide support for the atheistic belief maintained by the overwhelming majority of the NAS membership. And, those publications are the very ones the Dayton School District Policy mandates Mr. Scopes to use in his biology classes.

These publications provide a clarion call from the elite scientists of America for teaching 'scientific atheism' in the science classrooms of our public schools. These books are a not-very-subtle effort to require that the 'scientific truth' be taught that the universe and life itself is nothing other than the result of chance and accident. They make it very clear that any mention of planning or design should be banned from the classroom, for it carries religious overtones that are contrary to science. The

Preface to both of the NAS books that the School District requires Mr. Scopes to use in his biology class indicates that they were written to elaborate:

> '. . . the case against presenting religious concepts in science classes . . . and the reasons why only scientifically based explanations should be included in public school science courses.'

But these books then insist that the NAS belief in the ultimate reality of chance and accident must be included in classroom instruction because it is the only possible scientific explanation.

By proclaiming **without valid scientific evidence** that **undirected random** mutations are the basis for all adaptive evolution of living organisms and of life itself the NAS publications are simply stating a metaphysical-religious belief that has no place in science class.

Since the National Academy of Sciences is regarded as the highest scientific authority in America, their teachings are widely held to be most factual. Those teachings are regarded by most mainstream scientists and laymen alike to be 'scientific truths'. In these books the NAS far more than infers that atheism is a factual 'truth', they maintain that there is valid scientific evidence in support of the atheistic belief held by the overwhelming majority of the NAS membership.

Randomness / Accidentalness is a belief that may be true. Intelligent Design is a belief that also may be true. After all, ultimate reality either has to be purposeful or accidental – one or the other.

The doctrine of Intelligent Design is at core based on belief in a Purposeful God. If genetic mutations are not random but are somehow directed then ultimate reality is a Purposeful God.

The NAS doctrine of **randomness** is at core based on belief in scientific atheism's Accidental God. If genetic mutations are in fact random then ultimate reality is the Accidental God.

The United States District Court held in *Kitzmiller v. Dover Area School District* that purposeful ultimate reality (Intelligent Design) could not be taught in science class because it is a belief that is not falsifiable and for which there is no scientific evidence. Therefore Intelligent Design is not a valid part of science.

Intelligent Design has been banned from the science classroom, and rightly so, as a violation of the Establishment Clause of the First Amendment of this nation's Constitution:

> 'Congress shall make no law respecting an establishment of religion, or prohibiting the free exercise thereof . . . .'

Well, this trial is about the flip side of that coin. Accidental ultimate reality is a belief that is also not falsifiable and for which there is no scientific evidence. Therefore the doctrine of **randomness** is not a valid part of science. It also should be banned from the classroom in violation of the Establishment Clause.

Mr. Scopes fully agrees that religious concepts should not be presented in his science class. Mr. Scopes does not want to teach either the doctrine of Intelligent Design or the doctrine of randomness. Both doctrines cannot possibly be falsified. Both doctrines are metaphysical-religious beliefs for which there is no valid scientific evidence. Neither of those doctrines should have a place in Mr. Scopes' biology class.

The doctrine of randomness in living things is just another way of saying that natural selection is based on **random** genetic mutations. The explanations provided by the NAS insist that the evolution by natural selection of adaptive traits in living organisms must be done by **random** mutations. That explanation is not falsifiable and there is no scientific evidence in support of that explanation. We will prove that during the course of this trial.

If the doctrine of randomness is not supported by valid scientific evidence it clearly falls under the heading of a metaphysical-religious belief that must not be taught in the public school biology classroom. It must be banned from public school science class as a violation of the Establishment Clause."

Clarence Darrow asked permission to retrieve some further materials from the Plaintiff's table before continuing. The Judge said okay. The Plaintiff's attorney then continued.

"Your honor, this trial is not about Darwin's general theory of evolution.   Darwin described evolution as descent with modification through natural selection.   And, modern science has shown through abundant evidence that adaptive modification occurs through genetic DNA changes. Both parties to this lawsuit have stipulated that evolution occurs through changes (mutations) of genetic DNA.

Darwin's theory of evolution by natural selection is not at issue. We have no doubt of the validity of the theory. We have no doubt that adaptive change occurs over time through modifications of genetic DNA. What is at issue is the **mechanism** by which genetic mutations that result in adaptive change in living organisms occur.

This trial is about one word. The word is **random**. The word random is an adjective. It describes the noun that it precedes. The noun in this case is **mutation**.

The adjective **random** is defined by Webster's Ninth New Collegiate Dictionary as:

> 'Lacking a definite plan, purpose, or pattern . . . . Random stresses lack of definite aim, fixed goal, or regular procedure.'

Do genetic mutations that result in adaptive change occur at **random** or are they **directed** in some fashion that no scientist has ever been able to figure out? That is the question. And, there is no answer to that question that has been obtained by scientific evidence.

Yet, the NAS publications proclaim that the explanation of **randomness** has been clearly established by empirical scientific evidence. They proclaim that undirected **random** mutations is a scientific explanation. And they proclaim that this explanation must be taught in science class.

The National Academy of Sciences includes this statement in their 1999 book:

> 'No body of beliefs that has its origin in doctrinal material rather than scientific observation, interpretation,

and experimentation should be admissible as science in any science course. Incorporating the teaching of such doctrines into a science curriculum compromises the objectives of public education . . . . The growing role that science plays in modern life requires that science, and not religion, be taught in science class.'

The Plaintiff fully concurs with this position of the NAS. But, the inclusion of materials in their publications which explain that adaptive genetic mutations have been proven to be accidental and not directed is without scientific merit. That randomness explanation is not scientific. It is simply doctrine of a religion called 'scientific atheism'. That explanation has its origin in doctrinal material rather than scientific observation, interpretation, and experimentation. It is not a scientific explanation.

Whether genetic mutations that result in adaptive biological change are directed or random is not the stuff of some esoteric debate. It is most fundamental. The ultimate reality for directed change is a Purposeful God. The ultimate reality for random change is the Accidental God of scientific atheism.

When our children are taught in science class that scientific atheism is the 'scientific truth', the take home lesson for living life is the adoption of a philosophy called **nihilism** wherein:

- There is no basis for objective morality.
- There is no intrinsic value in human life.
- There is no purpose and meaning in life beyond that which we simply make up.

The religion of scientific atheism preaches this religious doctrine which, along with all other religious doctrines, should be expelled from the science classes of our public schools.

Both science and the law have come a long way since the *Scopes* trial of 1925. In 1925 John Scopes was on trial for violating a State Law that forbade a public school science teacher to:

> '. . . teach any theory that denies the story of the Divine Creation of man as taught in the Bible . . . .'

Since that time the United States Supreme Court has made it crystal clear that such religious instruction in public school science classes is prohibited by the Establishment Clause of the First Amendment of the United States Constitution. It is clearly unconstitutional to allow state-sponsored religious instruction to occur in public school science classes.

In the 1968 Supreme Court decision in the case of *Epperson v. Arkansas* Justice Fortas eloquently provided this instruction:

> 'Government in our democracy, state and national, must be neutral in matters of religious theory, doctrine, and practice. It may not be hostile to any religion or to the advocacy of no-religion; and it may not aid, foster, or promote one religion or religious theory against another or even against the militant opposite. The First Amendment mandates governmental neutrality between religion and religion, and between religion and nonreligion.'

As we gather here today everyone understands that a Purposeful God has no place in the science classroom. But, even as a Purposeful God has been expelled from the classroom, the Dayton School District has welcomed the Accidental God into the classroom with open arms.

Today, the state-sponsored religion being instructed in science class is not Christian creationism. Rather, it is the doctrine of scientific atheism. The randomness doctrine of scientific atheism preaches that life began and adaptively evolved through the natural selection of mutations that **just had to be random and not directed.**

Scientific atheism is clearly a metaphysical-religious belief not supported by scientific evidence. It has no place in science class. Mr. Scopes should not be forced to teach that unscientific, faith-based belief to youngsters in his high school biology classroom.

The Plaintiff will show through evidence produced in this trial that there is absolutely no scientific basis for teaching the doctrine of randomness in public school science class.

The *'Only Science in Science Class Policy'* of the Dayton School District forces Mr. Scopes to teach that unscientific belief. That

Policy is unconstitutional as a violation of the Establishment Clause and must be voided by this Court.

Thank you."

Judge Raulston gave Clarence Darrow the courtesy of allowing him to gather his papers and return to the Plaintiff's table and arrange himself there before continuing. He then recognized counsel for the Defense.

"Mr. Bryan you may proceed with your opening."

William Jennings Bryan seemed to spring out of his chair and began to speak even as he approached the lectern. He had no notes. He seemed to be a man in a hurry. He was by nature a cerebral attorney, highly skilled in courtroom drama. His height was average and he prided himself on his rail-thin appearance. He was a marathon runner and had the confidence obtained through twenty years of courtroom success.

Defendant's Opening Argument

"Your honor, to be honest, I am appalled by the elementary ploy being acted out today by Mr. Darrow. And, as you know I am more than a little disturbed by the fact that you did not grant our request for summary judgment to dismiss this ill-conceived case as being wholly without merit. I, of course, defer to your wisdom in letting this matter proceed at trial. But, I am sure that the trial will be a short one as we show beyond any doubt that all of the information and explanations provided by the NAS in their books are based on rock-solid science.

In the interest of brevity and justice, and in contrast to Mr. Darrow's ramblings about scientific atheism, I will make my opening statement short and sweet.

This case has nothing at all to do with religion. Quite the contrary, what the NAS provides in its books are scientific explanations based on scientific evidence. The NAS firmly believes that religion and science are separate and distinct practices. The practice of science **never, ever** considers religion. Indeed, in the

2008 NAS book *Science, Evolution, and Creationism* the authors go out of their way to give religion its due while separating religion from science:

> 'Science and religion are based on different aspects of human experience. In science, explanations *must* be based on evidence drawn from examining the natural world. Scientifically based observations or experiments that conflict with an explanation eventually *must* lead to modification or even abandonment of that explanation. Religious faith, in contrast, does not depend only on empirical evidence, is not necessarily modified in the face of conflicting evidence, and typically involves supernatural forces or entities. Because they are not a part of nature, supernatural entities cannot be investigated by science. In this sense, science and religion are separate and address aspects of human understanding in different ways. Attempts to pit science and religion against each other create controversy where none needs to exist.'

The fact that many NAS scientists do not believe in God has no bearing on their scientific inquiries. They are trained professionals, skilled in the use of the scientific method and they follow the scientific evidence wherever it leads them.

To contend that their commitment to pursue only natural explanations is some sort of religion of scientific atheism is poppycock. It is offensive and an affront to the professionalism of the esteemed scientists of the NAS.

Mr. Darrow and I have agreed on one thing that should be useful to disposing of this farcical case in a short time. Mr. Darrow has agreed that the only witness that he will call in this trial is the President of the National Academy of Sciences, Professor Noah Tall. We have stipulated that Dr. Tall speaks with the authority of the National Academy. As the newly-elected President of the NAS he will speak with expert authority on all scientific matters that Mr. Darrow wants to bring up. His answers to Mr. Darrow's questions will provide all the evidence needed to make a ruling for the Defendant in a very short while.

Your Honor, I am so certain that this matter can be resolved quickly and finally through the expert testimony of Professor Noah Tall that the Professor and I have agreed that I will, from this point forward and throughout his testimony, refrain from objections or cross-examination.

Professor Tall will be happy to answer all of Mr. Darrow's inane questions fully and scientifically. He needs no lawyer to defend the scientific truth that he can most-eloquently reveal. Many lawyers before Mr. Darrow have tried to question the truth of science. They have found that scientific truth is beyond doubt. Mr. Darrow will find the same. Mr. Darrow will soon discover that the truth of science will always prevail.

The Professor and I have agreed that Mr. Darrow can treat him as a hostile witness. He can lead the witness with inane questions till the cows come home. All we ask of Mr. Darrow and this Court is to allow Dr. Tall to answer the questions fully and completely. Given the opportunity for full explanation the scientific truth will always prevail.

Your Honor, I truly believe that as you hear the explanations and evidence provided by this most-esteemed scientist you may be inclined to shorten Mr. Darrow's opportunity to demean the NAS in particular and the dignity of science in general and provide a ruling for the Defense in a very short while. The truth of science is most compelling and cannot be confused, even by the legal brilliance of Mr. Darrow.

I promised to be brief, and I have been. That concludes my opening statement. Thank you."

* * *

With the conclusion of the opening arguments Judge Raulston called Dr. Noah Tall to the witness stand. Noah Tall was duly sworn-in as the only witness to be called in this bizarre case. Both the Plaintiff and the Defendant had indeed stipulated that Noah Tall was most qualified as an expert witness in science who could expertly testify in this trial to all things scientific.

Noah Tall's name belied his physical height. Yet, his five foot three inch height was no diminution at all to the stature and bearing of the man. He was strikingly handsome with dark wavy hair, including just a hint of gray in the temples.

I have always been amazed at the number of Marine Corps generals who are short in stature but grand in their physical command of the room. Noah Tall had the same bearing of a commanding general. An aura of authority and dignity surrounded him, as he was sworn-in to testify.

Clarence Darrow rose from the Plaintiff's table, secured his cane, walked to the front of the table and leaned back against it for some support. From that position he could both question the witness and effectively refer to notes on the table behind him.

* * *

Pardon again this author's digression. For purposes of clarity, yet brevity, the dialogue in the trial proper will be presented in a question and answer format denoted simply as Q and A. Each Q will represent a question from Mr. Clarence Darrow and each A will represent the answer provided by Professor Noah Tall, President of the National Academy of Sciences (NAS). End of digression.

* * *

Q: "Good morning Professor Tall. As you know I do have a lot of questions to ask you. I'd like to say at the outset that I have nothing but the highest esteem for you as a distinguished and dedicated scientist. I do not believe that I have ever had the privilege of questioning an expert witness as highly qualified as you are. My questions at times may seem to you to be naïve or elementary, but I assure you that my intention is none other than to discover the truth. If I come off as offensive or demeaning in any way I apologize in advance. That will never be my intent.

I'd like to start and conclude today's testimony by exploring the method by which science works. A clear understanding of that methodology should provide the necessary groundwork for my questioning during the remainder of this trial.

In the 2005 Pennsylvania case of *Kitzmiller v. Dover Area School District*, Judge Jones described the method of science in this manner:

> 'Methodological naturalism is sometimes known as the scientific method. Methodological naturalism is a "ground rule" of science today which requires scientists to seek explanations in the world around us based upon what we can observe, test, replicate, and verify.'

I'm sure that you are familiar with Judge Jones' ruling in that case that the theory of Intelligent Design could not be taught in public school because it could not measure up to the ground rules of science, what is called the scientific method.

Professor Tall, how would you describe the scientific method?"

A: "The scientific method is in essence a method of logic based on inductive reasoning. A scientist:
- observes a particular occurrence or set of occurrences in the natural world;
- forms a hypothesis based on those observations by his use of rational thought to explain how those particular occurrences happened;
- tests the hypothesis again and again to see if all such additional observations confirm the hypothesis.

The hypothesis is a scientific theory that provides a specific conclusion – a conclusive scientific explanation based on a limited set of observations. The empirical data that is accumulated through extensive testing and observations serves to strengthen the probable truth of the theory."

Q: "Judge Jones ruled that the Intelligent Design hypothesis was not a valid scientific theory and cannot be taught in science class. Do you agree with that?"

A: "I most certainly do. In order for a theory to be scientific it must be possible to actually test the hypothesis. In the case of the theory of Intelligent Design it is impossible to test the hypothesis. In order for a theory to be scientific it must be possible to show that the theory is wrong. The ability to be **falsified** is an essential feature of any truly scientific theory.

The necessity of falsification is why the scientific community has been so adamant in its opposition to teaching the theory of Intelligent Design in the biology classroom. It is impossible to falsify a theory that can provide an explanation for any observed biological phenomenon by saying that it was produced through the direction of an intelligent agent that cannot be observed. That explanation can never be falsified. That, therefore, can never be science."

Q: "So, purposeful intelligent direction can never be observed in nature?"

A: "No it cannot. When we observe in nature what seems to be purposeful intelligent direction we find that it has ultimately evolved from natural selection of random mutations of genetic DNA."

Q: "Is Darwin's theory of Evolution by Natural Selection a valid scientific theory?"

A: "It most certainly is. It is a theory that has been tested again and again and has always been verified. It is not the nature of science to ever provide absolute proof. Thousands and thousands of verifications cannot provide absolute proof for a theory, but one falsification can prove the theory wrong.

The strongest scientific theories have had the most verification and are considered by most scientists to be 'scientific truth'. The strongest of those theories include:
- cause and effect;
- a Sun-centered solar system;
- gravity.

And, the theory of Evolution by Natural Selection is now one of the strongest scientific theories, right up there with the theory of gravity."

Q: "Professor, we fully agree. We have stipulated that Evolution by Natural Selection is rightfully considered to be a 'scientific truth'. Evolution by Natural Selection says that living organisms evolve over time by inheriting adaptive traits that allow them to meet the challenges of their environment. The evolution of heritable adaptive traits is the result of changes (mutations) that occur in the information content that is contained in each organism's DNA. Science has proven all of this beyond any reasonable doubt. It is stipulated that Evolution by Natural Selection should be taught in science class.

The only question that remains concerning Evolution by Natural Selection is this. What is the change mechanism for modifying the information content of genetic DNA that provides for adaptive trait modifications?

- Intelligent Design; or
- Randomness.

The National Academy of Sciences explains quite clearly that the change mechanism is randomness. In your 1999 book *Science and Creationism* you provide this succinct and unequivocal explanation:

'Genetic mutations arise by chance.'

In your 2008 book *Science, Evolution and Creationism* you specifically explain that 'by chance' change mechanism in this manner:

'**Biological changes** that provide the raw material for evolution **are not directed** toward predetermined, specific goals. When DNA is being copied, **mistakes** in the copying process generate novel DNA sequences. These new sequences **act as evolutionary "experiments"**.'

Doesn't the National Academy of Sciences and nearly all biology textbooks now explain that all living organisms contain DNA and that the mechanism by which natural selection selects is **randomness – random** mutations of genetic DNA?"

A: "Yes. That is a fact."

Q: "How do you know that the mutations of genetic DNA are **random** mutations and not purposeful mutations?"

A: "Science deals only with the natural world, what Judge Jones called 'the world around us'. It leaves the realm of the supernatural to others. The only explanation possible in the natural world around us is randomness. A purposeful explanation is beyond the limits of the natural world around us."

Q: "Let me pursue this a little further to see if I really understand.
   You observe a bird building a nest for its offspring as it flies about in the woods and snuffles on the ground to acquire twigs and sticks and mud bits. The bird makes repeated trips to the emerging nest in the crook of tree branches and forms the twigs and sticks and mud into a bowl shape into which eggs will be laid. From your observations you form the hypothesis that a bird purposely builds a nest for its young. That hypothesis is tested again and again by further observations. Never is it observed that a bird builds a nest by random action, by accident. The hypothesis of purposeful nest building by birds is verified again and again. It therefore becomes a valid scientific hypothesis. Is that not purposeful intelligent direction that has been observed in nature?"

A: "No, it is definitely not. That is an example of instinctual behavior that has evolved over millions of years. The bird is simply following the instructions built into its DNA. Over many, many generations the genetic DNA was modified by random mutations that were selected by nature because those random mutations provided a survival advantage to the members of the species who possessed those modifications. The adaptive trait of nest building was the result of genetic DNA modifications. Random mutations served as evolutionary 'experiments'. The experiments that worked out the best were retained in the gene pool. No purpose and direction was involved at all."

Q: "So, let me make sure I correctly understand you. When a bird builds a nest it looks like it is engaged in purposeful action. That is what we observe. But, actually what it is doing resulted solely without direction or purpose. What we observe that seems to be purposeful action is actually the product of no direction and no purpose. Is that correct?"

A: "Yes."

Q: "How do you know that the modifications of genetic DNA that resulted in the evolution of this adaptive trait of nest building were in fact **random**?"

A: "We have been mapping and comparing the genetic makeup of many species, including our own, for many years now. The human genome project has been completed. We know the active protein structure of every part and particle of our species. We know what genes reside on each of our 23 pairs of chromosomes. And we have traced the evolutionary development of that genetic makeup by DNA developmental comparisons with other species, even to bacterial and animal beginnings. We have the physical evidence that shows how DNA changes have historically developed to provide evolving structure and biological systems."

Q: "But, when you make those comparisons and observations can you actually see that the changes that were made were **random**?"

A: "No. You cannot observe randomness in biology. But, we know that the adaptive changes were in fact random because that is the only explanation possible if we are to remain scientific. It is a valid inference that we make from confirmable empirical data."

Q: "What scientific criteria do you use to conclude that all genetic mutation is random?

A: "Mr. Darrow, the evidence for continuity in nature is overwhelming. The fossil record provides compelling evidence through radioisotope dating that there has most certainly been a progression of life through evolutionary time. The evidence shows that early eukaryotic creatures preceded fish; fish preceded amphibians; amphibians preceded reptiles; reptiles preceded mammals; pongid ape preceded hominid man. And, the fossil record contains evidence that clearly shows as these animals evolved over time, from fish to reptile to bird to human they continued to contain a similar fundamental anatomical structure. Although these structures look very different in form and may perform very different functions, the same basic bones are the same in porpoises and frogs and horses and bats and us.

Continuity in nature is compelling evidence of randomness. If genetic mutations were not random why would they evidence such continuity?"

Q: "Answering my question with a question does not provide scientific evidence to support your hypothesis of randomness. Evolution by natural selection of genetic mutations has been stipulated as true in this trial. That stipulation is well supported by the evidence of homology that you have just recapped. But, the existence of continuity in development provides no evidence whatsoever for your explanation of randomness. Our observations in everyday life would lead us to the exact opposite conclusion. Randomness never produces continuity.

I can think of another simple explanation for continuity in nature. The information contained in genetic DNA could have been initially programmed in a fashion that provided for just such continuity in evolution. But, of course, that explanation would infer intelligent design, and intelligent design just does not comport with science does it?"

A: "Again, Mr. Darrow, continuity is itself evidence for randomness. That is the only explanation possible if we are to remain scientific. You are free to offer snide remarks and believe anything that you want. I have told you the scientific answer."

Q: "Professor Tall, you have stressed that a scientific explanation must be capable of falsification if it is to be called scientific. How can you call the mechanism of random mutations scientific? Just like the mechanism of Intelligent Design, the mechanism of randomness is not ever capable of being falsified.

It is impossible to falsify a theory that provides an explanation for any observed biological phenomenon by saying that it was produced through a mechanism of randomness that cannot be observed. That explanation can never be falsified. That, therefore, can never be science. Isn't that so?"

A: "Randomness is simply the only explanation that can ever be science."

Q: "And, that scientific explanation tells us, in essence, that we must believe, based on your scientific authority, that **accidental** changes in the DNA world result in **purposeful** changes in the real world in which we live. Like the bird building its nest. Why would that make sense?"

A: "You may believe anything you want to. It may not make sense to you, Mr. Darrow. But that is the scientific truth."

Q: "DNA contains the ingredient that is absolutely essential for all life on this planet to exist, from the simplest single-celled organism to those of us in this courtroom today. That absolutely essential ingredient is information. Life is based on information. And, the information of life is contained in DNA.

Professor Tall, it is my hope in the days to come in this trial that we may all become more enlightened about how the DNA information of life has increased exponentially through the process of evolution by natural selection. Tomorrow let's begin by examining the nature and processing of the information contained in the DNA of all living things. Is that acceptable to you?"

A: "Mr. Darrow, I look forward to it."

CHAPTER 5

The Trial – The Second Day
The Nature and Processing of
Information in Living Things

Before beginning the second day it may prove helpful for the reader to understand an important asset being employed by Clarence Darrow, the Plaintiff's attorney, during the course of this trial. So, please excuse another digression.

Clarence Darrow was an experienced trial lawyer who was most accustomed to jury trials. By common practice he employed a trial psychologist to assist him in the *voir dire* process of selecting a jury and to help him gauge the reactions of the jurors throughout the trial. The skilled attorney understood that the psychologist had developed skills of 'reading' the reactions of potential and seated jurors far better than he could.

But, remember that these proceedings are taking place as a bench trial, without the benefit of a jury. In this trial Judge Raulston is the sole trier of facts and neither Clarence Darrow nor any trial psychologist that he had previously employed had ever before encountered Judge Raulston.

Clarence Darrow figured that it would be most useful to employ someone who could correctly 'read' the reactions of Judge Raulston. He decided to pursue the employment of not a psychologist but, rather, someone who had observed Judge Raulston in his courtroom a lot. So, discrete inquiries around the courthouse in the weeks before the trial produced his acquaintance with Abraham Spyer. Since childhood, Abraham was regarded by all who knew him to be a person so completely invested with the attribute of honesty that his given name had been long forgotten. He was known only as 'Goody' Spyer.

Goody Spyer was a retired high school basketball coach. He had moved to Dayton about five years ago and had become fascinated by his new-found avocation – courtroom watching. After his retirement Goody Spyer had become a seasoned courtroom watcher. He found that the real thing was much more interesting than watching *Judge Judy* or *The People's Court* on TV.

As luck would have it for Clarence Darrow, Goody Spyer had chosen John Raulston's federal courtroom as his home base. He had observed Judge Raulston in action ever since the new courthouse had opened in Dayton three years before. If anyone could give Clarence Darrow a good 'read' on Judge Raulston during the trial it was Goody Spyer. So, Goody Spyer was thus employed by Clarence Darrow.

Clarence Darrow's methodology for gaining the most from Goody Spyer's observations was a simple one. Before each day of trial the Plaintiff's attorney would provide to his seasoned Judge-observer a briefing book containing the main points that he wanted to successfully convey through the course of that day's questioning of the trial's expert witness, Noah Tall.

Then, upon receipt of a secret signal from Goody Spyer during the course of each day's proceedings, Clarence Darrow would request a recess. The secret signal would be received as Clarence Darrow observed Goody Spyer artfully arranging a challenging comb-over in an attempt to conceal his balding head.

The recess would allow him the opportunity to gain critical feedback from Goody Spyer. Were the questions and answers proving effective in making his case in the eyes of Judge Raulston? What needed further clarification? What needed additional stress? Thereby, Clarence Darrow would rely heavily on the observations of Goody Spyer in honing the presentation of his case with mid-stream adjustments. After the recess ended he would be prepared to fill in the gaps.

Well, as luck would have it with me, Goody Spyer had been my high school basketball coach in Hazel. I knew and liked Goody a lot. And, apparently, the feeling was mutual. That fortuitous circumstance will allow the reader to gain insight into Clarence Darrow's consultations with Goody Spyer during the course of this trial. End of digression.

* * *

After the bailiff's opening ritual, the second day of the trial began with Noah Tall again taking his seat on the witness stand. Judge Raulston conferred privately for several minutes with Clarence Darrow and William Jennings Bryan in front of his bench. Clarence Darrow then took his place in front of the Plaintiff's table, leaned back, and began.

* * *

Again, for purposes of clarity, yet brevity, the dialogue in the trial proper will be presented in a question and answer format denoted simply as Q and A. Each Q will represent a question from Mr. Clarence Darrow and each A will represent the answer provided by Professor Noah Tall.

* * *

Q: "Good morning Professor Tall. I hope you had a restful evening. We have quite a bit of ground to cover this morning.
   I'd like to begin today to learn more about DNA itself. In yesterday's testimony you confirmed that all living things that have ever lived on Earth contain DNA. What exactly is DNA?"

A: "DNA is deoxyribonucleic acid. It is a genetic nucleic acid. There are actually two types of genetic nucleic acids, RNA and DNA. The DNA molecules reside in the cell's nucleus and supply the blueprint and recipe for cell development. The RNA molecules are the messengers (mRNA) that convey the proper blueprint information to the specific location in the cell where the protein machine is being assembled (ribosomes) and then translate and transfer (tRNA) the information into the language of amino acids necessary to build each protein in accordance with the specific sequence of amino acids originally dictated by the DNA in the first place.
   The RNA and DNA molecules are linear chain molecules called polymers. These nucleic acid polymers are long chain sequences of nucleotides.

The DNA molecule has two strands of nucleotides in the shape of a double-helix, formed by the complementary base pairing of the nitrogenous bases of the nucleotides. Each turn of the helix contains ten nucleotides. The mRNA molecule has a single linear strand of nucleotides. The tRNA molecule has a complex 3-D structure that arises from complementary base pairing. The unique structures of these nucleic acids enable them to precisely bond in the nature required to provide instructions to amino acids for the formation of each protein of each and every living organism.

In short, the nucleotides string together to make up a strand of DNA called a gene that instructs the construction of a protein."

Q: "Professor, if I could interrupt for a moment so I don't get lost. How many nucleotides string together to make up a strand of DNA called a gene?"

A: "The number of nucleotides chaining together to provide the information necessary for the construction of each individual protein varies greatly. For example, the gene for the protein insulin is just over 200 nucleotides long, the gene for the protein hemoglobin is about 400 nucleotides long, while myosin, the gene for the protein that controls bodily muscles is about 4,000 nucleotides long."

How DNA Constructs Proteins

Q: "Thank you. Would you please continue with your explanation of the workings of DNA?"

A: "Of course. The synthesis of proteins by a cell can be seen as the result of the partnership of one group of bio-molecules (amino acids) with another group of bio-molecules (RNA and DNA nucleic acids). The nucleic acids supply the blueprint and recipe and extract the information from the blueprint at the just-right location in the cell for protein construction. Then, through a series of exquisite transformations, the amino acid building blocks fulfill the construction project by synthesizing each protein needed by the cell in order for it to properly function.

The blueprint for all information about each and every part of each living organism on this planet is contained in the DNA of that organism. DNA is the repository of all life information and contains both the blueprint for the construction of each part of every living organism, and the recipe for how to bring the blueprint to life.

DNA is comprised of building blocks called nucleotides. Each nucleotide contains three specific parts:
- a 5-carbon sugar group;
- a phosphate group; and
- a nitrogenous base.

The sugar group and the phosphate group chain together with a linear chemical bonding that forms the nucleotide backbone of the DNA. The nature of their chemical bonding results in a 3-carbon atom at one end of the nucleotide chain and a 5-carbon atom at the other end of the chain, thereby making the nucleotide chain electrically charged in order to electrically and chemically bond with the next nucleotide sequence in the DNA chain.

Attached to the side of each alternating sugar group and phosphate group is a distinctive nitrogenous base. The nitrogenous base is the only functional unit of the nucleotide and is, therefore, the defining feature of each segment of DNA constructed by the nucleotides. There are only four nitrogenous bases and they are configured along opposite sides of DNA strands as complementary base pairs. The four bases are: T (thymine); A (adenine); C (cytosine); and G (guanine). A always bonds with T and C always bonds with G.

The alternating base pairs are strung out in linear sequences along long strands of DNA. The nucleotides making-up the DNA are wound in a right-hand direction into complementary strands to form the famous double-helix. And, the DNA strands are always anti-parallel running in opposite directions, thereby becoming polarized."

* * *

Professor Noah Tall paused a few moments to gather his thoughts and take a drink before continuing.

* * *

"DNA is housed in chromosomes. The chromosomes are made-up of both DNA and various proteins. A chromosome contains 5 to 10 times more protein than DNA.

However, only the DNA portion of the chromosome contains the blueprint and recipe information. The specific segment of a DNA sequence that holds the information required to express a specific protein is called a gene.

The blueprint and recipe information contained in the genes along the strands of DNA is in a code. The code is a simple yet elaborate structure. Once the code is translated, the message simply specifies the sequence of amino acids necessary for the construction of a single protein to be used by the cell. Again, the coded information contained in DNA is simply the sequence of amino acids to be used in the construction of proteins.

That's the nutshell version of what DNA is and how it works."

DNA Information is in Code

Q: "There sure is a lot of meat in that nutshell. You have mentioned on several occasions the coded information in DNA. What is this DNA code?"

A: "The historical central dogma of genetics, which we now know is an overly-simplistic explanation, informs us that a gene is simply a specific segment of DNA that contains the information required to construct a specific protein. That information directs the specific sequence of amino acids necessary to construct a particular protein. But, the nucleotides defining the genes of DNA cannot talk directly to the amino acids that make-up proteins. A code is needed. Each gene codes for a protein.

Nucleotides that chain together to make-up DNA use a language containing but four base pairs of nitrogenous bases (remember A always pairs with T and G always pairs with C). Living proteins contain amino acid sequences always comprised of some of twenty specific amino acids. A code is needed to translate the language of DNA nucleotides (four units) into the language of amino acids (twenty units).

It took Francis Crick and James Watson many years to first discover the amazing properties of the DNA double-helix and then to discover the three-letter genetic code necessary to translate the language of DNA and RNA nucleotides into the language of amino acids in order to construct proteins.

The genetic code is a three-letter code using specific combinations of three nucleotide bases called codons. Crick found that the three-letter code needed to uniquely identify sequences of the four possible bases (4x4x4=64) provided 64 combinations.

Each unique combination of three nucleotide bases is called a codon. Each codon codes only for a single amino acid. Sixty of the possible 64 codons code for a specific amino acid. Three of the codons code only for a stop signal. One of the codons codes as a start signal and thus provides an unambiguous marker for the first nucleotide base in a sequence of codons in mRNA.

In short, DNA contains a quite complex coded language structure that is used to transmit the information necessary for the construction of all of the proteins used by living cells that are required for the living organism's functions."

DNA Specifies the Order of Proteins

Q: "How does a cell know which proteins need to be constructed within it in order for it to properly function?"

A: "DNA provides the instructions for the construction of what are called 'transcription factors' that are themselves specialized proteins that are used for 'gene expression'. The 'transcription factors' assist DNA in determining the specific proteins the cell needs to construct. But, for simplicity's sake I better work up to 'transcription factors'.

Okay. The DNA library contains all of the information for constructing each and every protein in each and every cell in the body. Protein construction leads to functional operation for each and every part of the body and all the connections therein. The DNA library reposes intact in the nucleus of each and every cell of the body.

In order for the DNA information to be useful in constructing the protein needed by a cell it must be extracted from the DNA. Each living cell will construct only those proteins whose functions are expressed by the segments of the DNA particular to that cell's proteins. Each particular segment of DNA is called a gene. You have heard the term gene expression. Well, that's what it means.

A single human cell may contain thousands of different proteins. So, thousands of DNA segments (genes) will need to be expressed in order to complete that particular cell's development. Thousands of genes need to be expressed to make thousands of different proteins in a single cell.

Okay. The chromosomes reside in the cell's nucleus. The DNA for the organism resides on the chromosomes. There is a specific segment of the DNA, called a gene, that contains the specific information necessary to construct a specific protein needed by the cell. How does the particular cell know which proteins it needs? Here's how.

DNA itself tells the cell which proteins it needs by specifying the construction of 'transcription factors'. The beginning sequence of nucleotides on a gene is a 'start' signal. So, the DNA gives instructions to the cell for not only which genes to copy but where to begin copying. The specialized proteins called 'transcription factors' attach to the DNA segment (gene) at just the right spot and issue a start signal. At the start signal on the gene a protein enzyme called RNA polymerase attaches at the just-right location on the DNA and begins to unwind the two strands of DNA. RNA polymerase then begins the transcription process. Transcription is correctly understood as DNA-directed RNA synthesis.

Remember, the functional role of mRNA is to simply act as a messenger after it transcribes DNA's language into a form that can be accurately handed-off to tRNA at the protein building site in the cell (the ribosomes).

RNA polymerase then proceeds to copy the entire gene segment of one of the DNA strands and thereby transcribe it into RNA format. Only one strand of the DNA gene segment is copied into RNA form because the exact sequencing nature of base-pairing requires only one strand.

After a 'rough copy' has been transcribed, a number of specific protein enzymes attach to the 'rough copy' and proceed to edit out a large sequence of the information that has been transcribed. Amazingly, as much as 90% of the transcription is at this point removed and discarded. The 10% remainder is spliced back together to make a much-smaller mRNA strand.

Once copied, edited-out, and spliced back together the new strand of RNA is snipped-off at a 'stop' signal which is designated by DNA as the last series of nucleotides on a gene. The DNA strand then rewinds to its original double-helix form once the RNA strand is snipped-off.

While still in the nucleus, the final editing process for the mRNA strand takes place. The ends of the strand are chemically modified by other enzymes. A guanine (G) cap is added to one polarized end and an adenine (A) tail is added to the other end. We now have a 'mature' strand of mRNA.

At this point the mature mRNA strand, containing the encoded information necessary to construct the protein, physically departs the cell's nucleus. The mature mRNA strand travels to the correct protein construction site in the cell's cytoplasm (the ribosome) and hands off its work to its teammate, tRNA. I've covered that process in my earlier testimony, so I won't repeat myself.

In short, that is how 'transcription factors' work with DNA to construct all of the proteins necessary for a specific cell's functions."

Q: "That is an incredibly complex process. It is absolutely amazing. But, as you have been describing the process all sorts of 'chicken and egg' questions come to mind. One in particular is most compelling.

How did the DNA segment (gene) that provided the instructions for the construction of a 'transcription factor' protein know where to begin construction before the 'transcription factor' necessary for the expression of that gene was itself evolved?"

A: "Mr. Darrow, that is actually a perceptive question. Science cannot at this time provide detailed answers for the evolution of each step of the process. Suffice it to say that each developed through quite natural processes in accord with evolution by natural selection of random mutations of genetic DNA."

* * *

If Clarence Darrow was playing poker he should have folded his hand. His frustration with Professor Tall's pat randomness answer was becoming more and more visible. And, Noah Tall seemed to be enjoying that. The Plaintiff's attorney noticed Goody Spyer's comb-over action out of the corner of his eye, and the Judge granted his request for a recess.

Goody advised that it would be a good idea for Clarence Darrow to take a break from all the technicalities and try to focus on the larger picture of the 'smart stuff' in DNA.

* * *

DNA is Smart Information

Q: "Okay, Professor. Let's return a bit to that DNA code. You indicated that DNA is itself a complex coded language. Other than DNA, can you give me an example of any other coded language, anywhere on Earth, that developed unintentionally, without direction or purpose?"

A: "No, I cannot."

Q: "Do you believe that DNA is a coded language that developed unintentionally, without direction or purpose?"

A: "Yes. That is the only explanation that is scientific."

Q: "And, by scientific, you mean that it must have derived from a natural, undirected and non-purposeful cause. Is that correct?"

A: "We have covered the same subject many times. The answer is still yes."

Q: "Do you think that DNA itself is intelligent information or smart information?"

A: "I don't understand your question."

Q: "Is DNA smart? Does DNA think and figure out solutions to problems?"

A: "No. Of course not. DNA is simply coded genetic information. It is no more intelligent or smart than the words on a piece of paper can be termed intelligent or smart."

Q: "Well, DNA causes things to happen, but the words on a piece of paper can't cause anything to happen. Can the words in a book that provide the precise detailed instructions for how to construct a table actually cause the table to be built?"

A: "Of course not."

Q: "But, let's look at the same thing with DNA. The coded words contained in DNA not only provide the precise detailed instructions for how to construct and maintain a complete living organism, the coded words also activate the information and bring the organism to life. Doesn't DNA 'thought' alone result in action and isn't that a prime example of intelligent or smart information?"

A: "No." That's a bit of unscientific romanticism."

Q: "Professor, let me read this passage from a book that I have recently read concerning how DNA operates, and then ask you a specific question. The passage uses our own species as the pointed example:

> 'DNA starts off and is contained in one single, solitary cell, a fertilized egg. That initial DNA molecule is composed of 23 pairs of chromosomes – ½ from our mother and ½ from our father. That initial DNA molecule contains all the information required to construct and grow and protect and maintain a unique living, thinking human being for an average lifetime in America of close to eighty years. That single molecule of DNA contains all of the information that has ever been written in books about human anatomy and physiology, and much, much more that is still missing from the library of human knowledge.
>
> But, that is but a small part of the wonder of DNA. DNA 'thought' alone then proceeds to transform the information within into physical reality. The DNA in the initial single cell assembles the raw materials of nucleic acids and amino acids within the cell and then proceeds to use them to construct a number of embryonic stem cells capable of becoming most anything. It then proceeds to differentiate further cells with further and further and further specificity into a blood circulation system, neural electrical circuit, heart, lungs, kidneys, stomach, hands, fingers, feet, toes, eyes, ears, lips . . . . Nine months later we hear a newborn baby's first cry and see the first smile (probably gas). Then, growth continues through infancy and childhood and the teenage years until mature adulthood is finally reached. At each stage we are protected from disease by an elaborate immune system and an exquisite blood coagulation system. And, each part and each system was constructed strictly in accord with the information contained in our DNA.

The resultant adult human being now contains literally trillions of living cells. Each of those cells contains the same exact DNA molecule. Each of those cells (which contains the same DNA molecule) knows precisely what genes need to be 'expressed' from the proper segments of the DNA molecule in order to provide for constructing the appropriate proteins necessary for the exact action needed for the proper function of that particular cell. In each and every cell DNA not only 'knows' what raw materials to use in the construction project, it causes the construction to occur and to be maintained as part of a fully-functioning living organism.

Not a single human thought is required to accomplish all of this. The only 'thought' is that of DNA itself. And, DNA itself puts 'thought' into action.'

Now, my question, Professor Tall, is this. From that description, what conclusion can possibly be drawn except that DNA is intelligent or smart information?"

A: "A romantic description does not make DNA anything other than what it is. DNA is coded information, nothing more and nothing less."

Where does the Coded Information in DNA come from?

Q: "Where does coded information come from? For that matter where does any information come from?"

A: "Everything contains information. All energy and all matter contain information. Water supplies us with the information that it contains hydrogen and oxygen. You touch a hot stove and it provides you with the information that it is hot. Everything contains information and provides us with information."

Q: "But, is any of that information useful without intelligent input?"

A: "The information contained in DNA may not be useful for our scientific research without our intelligence. But, it is certainly useful to living organisms who are built and operated in accord with the information contained in their DNA. And, no intelligent input is required for that. The coded information in DNA simply exists. Nothing more.

    The laws of physics tell us how the electrical bonding of five natural elements combine to create an organic molecule of DNA. Because we understand physical laws we can understand the information contained in DNA. That provides us with knowledge."

Q: "The laws of physics explain the electrical bonding. But, do the laws of physics require the long coded sequences to occur in the specific order that they do?"

A: "No. They are guided by the laws of biology, specifically, evolution by natural selection of random mutations."

Q: "Professor, there is no scientific basis whatsoever for that answer. What part of the law of evolution by natural selection can scientifically explain how the specific ordering of nucleotides necessary for functional DNA came about naturally?"

A: "Mr. Darrow, again, I don't know how many times I have to repeat myself. The natural process that produces the information in DNA required for life is random mutation."

Q: "So, again, your answer is random mutations. Let me understand clearly. Random mutations produced all the specific sequencing of nucleotides to provide the information, contained in a coded language, that must be translated from the language of nucleotides into the language of amino acids and then used to assemble all of the proteins and cells of every living organism. And, all that happened without direction or purpose, by random mutations?"

A: "Correct. That is the only acceptable scientific explanation. Anything else is simply unscientific."

Q: "The nucleotide sequences in DNA follow the laws of physics. But the laws of physics do not necessitate the specific arrangements of those nucleotides. In non-living matter the laws of physics necessitate elemental interactions. In living matter they do not.

Yet, the information contained in the non-necessitated nucleotide arrangements of DNA is phenomenal, indeed beyond human comprehension. Noted astronomer Carl Sagan explained that the amount of information stored in the DNA of a single-celled bacterium would fill 100 printed pages of a book. The information stored in the genes of a single-celled amoeba would fill eighty 500-page books. The information contained in the DNA within a single human cell would fill a library of some 1,000 books, each 500 pages in length. That is the amount of information contained in each and every one of our 100 trillion cells.

The human genome is the list of coded instructions necessary to make-up a human being. There are over three billion letters (base pairs) in the human DNA code. Those coded letters reside in the nucleus of each and every one of the trillions of cells that make-up the human organism. That information is tightly wound in the pattern of a double-helix on twenty-three pairs of chromosomes inside each cell nucleus. If the DNA information reposing within a single cell were unwound it would straighten-out to a length of about five or six feet. If all the DNA information in a single human being were straightened-out and placed end-to-end it would reach from the Earth to far beyond the Sun. That amount of information is, indeed, beyond human comprehension.

My question, again, Professor. Is it your explanation that all of that specific information was produced without any direction or purpose by random mutations of genetic DNA?"

A: "Mr. Darrow, you already know my answer. Yes. There was no direction and no purpose. All occurred naturally. Random mutation is the only natural explanation that is acceptable in science.

Information is simply a feature of the energy and matter in the natural world."

Q: "Where does the information in DNA come from?"

A: "This is getting tiresome. DNA is simply a form of matter. And information is a feature of all matter. The commonality of the DNA information code and the commonality of the amino acids used by DNA to construct proteins is itself evidence of the fact that the basic processes of life are most highly conserved. As we explain in *Science and Creationism*:

> 'The code used to translate nucleotide sequences into amino acid sequences is essentially the same in all organisms. Moreover, proteins in all organisms are invariably composed of the same set of 20 amino acids. This unity of composition and function is a powerful argument in favor of the common descent of the most diverse organisms.'

Can't you understand, Mr. Darrow, the information needed for life is simply contained in the matter of the nucleotides that make up the matter we call DNA?"

Q: "In the name of sanity, let me end this circular reasoning by asking this question. You have explained that all energy and all matter contain information. Is that information useful to us without our intelligent input?"

A: "In order for information to be useful to us we must be able to use and manipulate the information. That requires the use of our intelligence."

Q: "Isn't that precisely what is happening within the cellular structure of each living organism on Earth? Isn't the cellular process that you have described for us an excellent example of understanding and using and manipulating the information residing in DNA?"

A: "No. I would not say that. What I would say is that the living organism simply follows the information contained in DNA by rote."

Q: "Doesn't that tell you that intelligence is required in order for DNA to contain information that it then uses and manipulates to assemble and maintain living organisms?"

A: "No. That simply tells me that you are not a scientist. DNA is simply a molecule comprised of nucleotides. Nothing more and nothing less."

Q: "Where did DNA come from?"

A: "As I have already explained, DNA forms from five nucleotides. Each nucleotide contains three specific parts:
- a 5-carbon sugar group;
- a phosphate group; and
- a nitrogenous base."

Q: "You testified earlier that this same DNA molecular structure is found in all living things. So, DNA itself seems to be very highly conserved. It has not evolved into something else has it?"

A: "No. The DNA structure has not evolved. The amount of information contained in a DNA molecule has greatly increased as life has evolved, especially intelligent life. But, the basic structure and sequencing mechanisms have not changed."

Q: "Professor, which of the following explanations regarding the information contained in DNA is the more scientific to present to youngsters in public school science class?

- The information in DNA originated from mutations of simpler non-living molecules that were undirected and random.
- No one knows how the information contained in DNA originated."

A: "DNA evolved from mutations of simpler non-living molecules. Those mutations were undirected and random. That is the only natural answer. And, that is the only acceptable answer in science class."

* * *

With that answer from Noah Tall, Clarence Darrow thanked him for his testimony and Judge Raulston adjourned the Court for the day.

CHAPTER 6

The Trial – The Third Day
Evolution from First Life to Primitive Animals

The exchanges of the previous day between Clarence Darrow and Noah Tall would prove to be but playful banter compared to the dialogue that would flow from the questions and answers regarding the subject of the third day. Again, for purposes of brevity, yet clarity, the dialogue will be presented in a question and answer format, beginning with the first question from Mr. Darrow.

* * *

Q: "Good morning Professor Tall. I trust that you had a pleasant evening. Yesterday we examined how all living things require the information contained in their DNA in order to live. Today I'd like to examine the process by which living things actually evolve through changes in that DNA information.

Could you begin that examination by providing a simple explanation of how Darwin's Theory of Evolution by Natural Selection actually works? You know, the theory of evolution that we have already stipulated as being true. Could you help us with that?"

A: "Of course. Charles Darwin's theory of evolution basically said that living organisms changed over time by descent with modification. The specific term that he used was 'descent with modification'. Organisms who possessed the traits that made them more able to survive would pass on those traits to offspring. Such favorable traits became a part of the organism's gene pool and became predominate for the species. Nature in essence 'selected' the traits that modified the species. That is known as natural selection.

Darwin's theory is solid science. It has been updated to include details that were not available to Darwin in the

19th century. For instance, Darwin was quite certain that adaptive trait modifications that resulted in the evolution of species were the result of nature selecting from a random pool of candidate modifications. He did not yet know how, but he was quite certain that the selection was undirected. Today, with the knowledge that science now has regarding DNA and species genomes, we know that the candidate pool of modifications for adaptive change results from either DNA genetic copying errors or an alteration of genetic DNA in response to some radioactivity or cosmic rays or some poison in the environment. All adaptive change is, therefore, the result of random genetic mutation."

Q: "You have added quite a bit to the stipulations we agreed to. The stipulations did not include the assertion that genetic mutations are undirected and random. That is the very issue to be determined by this Court, is it not?"

A: "It may be an issue for you Mr. Darrow, but not for me. Science is quite sure that genetic mutations are undirected and random."

Q: "Well it certainly is the issue before this Court, otherwise Judge Raulston would not be conducting this trial. We will examine that very issue in much greater detail as we proceed. So, let's do just that and get to the next question.

Darwin's theory proposed, and we have stipulated this point as scientific fact, that all living organisms on Earth could ultimately trace their ancestry back in time for billions of years to a forbear in the form of an organism that had but one single cell. Isn't that true?"

A: "Yes."

It All Started With Bacteria

Q: "Wasn't that first life some kind of bacteria?"

A: "Technically, it is called archaebacteria. But, to keep it simple, we'll just call it bacteria."

Q: "What are the essential features that even these most primitive organisms must possess in order to 'be alive'?"

A: "In order to be alive single-celled living organisms had to be capable of doing a minimum of three things. Each bacterium has to:
- store and process information;
- acquire and use energy; and
- reproduce its cells and itself."

Q: "What is single-celled bacteria made of?"

A: "First of all, a bacteria cell contains the nucleic acids, RNA and DNA. These nucleic acids provide the information necessary for the construction of the proteins that perform the functions of the organism. Without the information contained in the DNA and transcribed by RNA into amino acid language the organism could not make the proteins required for all the cell's functions.

After RNA transcribes the DNA information it transfers the transcribed information to the ribosome. The ribosome is the cell structure that serves to translate the nucleic acid language and then to build each protein needed by the cell by assembling amino acids into proteins in accordance with the instructions contained in the DNA information code. We covered all this yesterday."

Q: "And all that transcribing and transferring and translating of the information in DNA takes place without any direction or purpose?"

A: "Correct."

Q: "So, even single-celled bacteria have this DNA that provides the information needed for cell development?"

A: "That's true. All living organisms on Earth have the information they need for life contained in their DNA. Mr. Darrow, we have already covered this pretty completely."

* * *

At this point Judge Raulston asked both Clarence Darrow and William Jennings Bryan to join him in a side bar before his bench. The Judge wanted to provide William Jennings Bryan the opportunity to change his mind and to begin to engage in objections to leading and hostile questions from Clarence Darrow to Noah Tall. The defense counsel declined, reiterating his statement that Clarence Darrow could treat Noah Tall as a hostile witness and ask leading questions as long as he wanted. All he asked was that Noah Tall be given the opportunity to answer all questions fully and completely. He insisted on retaining his position that, given the opportunity for full explanation, the scientific truth would always prevail. With that reiteration Judge Raulston seemed satisfied and asked Clarence Darrow to continue.

* * *

Evolution Requires Changes (Mutations) In DNA

Q: "So, the first living organisms on Earth, single-celled bacteria, had the information they needed for life encoded in their DNA.

Dr. Tall, it is my understanding from reading your books that evolution proceeded by nature selecting random mutations of DNA that provided a survival advantage to the organism. Were all evolutionary changes actually selected from DNA copying mistakes made during the reproduction process or from accidental DNA damage resulting from some cosmic radiation or some other poison in the environment?"

A: "Yes. That is basically correct."

Q: "So, all adaptive evolutionary change then proceeded by nature selecting from random mutations of genetic DNA?"

A: "Again, yes."

Q: "Would you please explain the sequence through which living organisms began to evolve from that first bacterial life by the use of that process?"

A: "Over the course of about 500 million years some bacteria clumped together into large masses. Today we find on Earth evidence of that in the form of what are called stromatolites. They are sedimentary fossilized layers of blue-green algae bacteria. The fossil record dates them to about three billion years ago, or about ½ billion years after first life. Those blue-green algae bacteria had by that time learned how to extract the energy they needed for life directly from sunlight through the process we call photosynthesis."

Q: "I know that we *Homo sapiens* are creatures who have to eat in order to stay alive. I assume that bacteria have to eat something in order to stay alive. If that is true, what do they eat?"

A: "Just like us, all animals need to eat other living things or things that used to be alive in order to obtain the energy they need for life and living. By contrast, the first bacteria and many bacteria species as well as green plants today are autotrophs. They extract the energy they need to live directly from their environment. The first bacteria basically got their energy from eating rocks or gases like sulfur and methane. Then, as I just explained, certain bacteria, such as blue-green algae bacteria, evolved the elegant process of photosynthesis. Green plants perfected that process."

How Photosynthesis Developed

Q: "Would you please give us a thumbnail sketch of this elegant process of photosynthesis?"

A: "Sure. I'll explain the process that green plants use. Photosynthesizing bacteria use a very similar process that differs in details but not in concept.

First bacteria and then green plants figured out a way to capture the energy in the photons of light from the Sun that they absorb and convert into useful energy without it being lost as heat. This light-energy-capture occurs within the cellular structure of the leaves of green plants.

The cells of plant leaves (plant cells, never animal cells) contain distinctive organelles called chloroplasts. The chloroplasts have their own membrane system, and, embedded in the membranes are lots and lots of chlorophyll and other pigment molecules. Chlorophyll is used as the primary light gatherer in all plants and algae and in certain photosynthetic bacteria called cyanobacteria. Those are the same blue-green algae bacteria that massed together to form the stromatolites that we still find on Earth today. The specific job of chlorophyll molecules is to capture the energy of sunlight. Here's how they do it.

When a photon of sunlight encounters and hits a molecule of chlorophyll it excites the electrons in that molecule to a higher energy state. In plant leaves the molecules of chlorophyll are tightly packed together into what is called a photo system. So, when a photon strikes a molecule of chlorophyll the electron that it excites does not quickly decay to a lower energy level and is not fluoresced back into space at a longer wavelength. Rather, the chlorophyll molecule passes the photon on intact to the next molecule in the photo system, and so on and so on. When the photon reaches the chlorophyll molecule next to the primary electron acceptor molecule in the center of the photo system a marvelous transition occurs. The excited electron in that adjacent chlorophyll molecule is transferred intact into the electron acceptor molecule. The chlorophyll molecule loses an electron (losing an electron is called oxidation). The electron acceptor molecule gains an electron (gaining an electron, while counter-intuitive, is called reduction). This intricate process is called oxidation-reduction, abbreviated as a redox reaction.

Once a redox reaction occurs, once the energy of a photon causes an excited electron to leave the chlorophyll molecule and be added to the electron acceptor molecule, the captured energy is stored in a more lasting state. Now it must be converted into a more usable form of energy and stored for future use. Here's how that happens.

Immediately after the electron acceptor molecule gains an extra electron (is reduced) it transfers that electron to an electron transport chain embedded in one of the chloroplast's membranes. The energy level of the added electron is gradually stepped down through a series of redox reactions as it proceeds along the electron transport chain. The energy released from the electron in these reactions is used to pump hydrogen ions against their concentration gradient across the chloroplast's membranes. The energy, stored in the concentration gradient of hydrogen ions, is used to drive the production of the ubiquitous energy molecule called ATP, used by all living organisms, through the help of a protein enzyme called ATP synthase.

Ordinary water is a necessary reactant in photosynthesis. Electrons from water fill the holes created through the redox reactions and donate electrons to build high-energy glucose molecules and produce oxygen as a waste product.

The chemical reactions through the electron transport chain result in the production of a small amount of ATP and energy retained in the electron transport molecules themselves. However, these forms of energy do not have a very long 'shelf life'. ATP has to be used fairly quickly once it is produced, or it decays into inorganic ADP. Therefore, green plants had to figure out how to store this captured energy in a longer lasting form of high-energy organic molecules that can be used later. That is performed in a three-phase process, called the Calvin cycle. The Calvin cycle is brought into play in order to use the initial organic energy source to now synthesize the high-energy sugar called glucose.

Photosynthesis is quite a complicated and elegant process, but that is a thumbnail sketch. The blue-green algae and green plants extract carbon dioxide and water from their environment and combine them with sunlight to produce a food source for themselves and for us. All of us mammals and other animals evolved after bacteria and green plants. We are unable to extract energy directly from our environment. So, we owe our very existence to the evolution of photosynthesis by bacteria and green plants."

Q: "To state it mildly, that is simply amazing. And, you believe that the bacteria that first developed photosynthesis did so in order to meet an environmental challenge? What would the environmental challenge have been that resulted in this evolutionary change developed by bacteria?"

A: "The iron, or sulfur or methane that the ancestor bacteria used as an energy source was likely challenged."

Q: "But, those iron and sulfur and methane eating bacteria continued to exist, in fact they continue to exist on Earth today. They did not eat each other and their food source has lasted for billions of years, even to this day. Why would such a complicated process as photosynthesis evolve by undirected random mutation of bacterial DNA in order to solve an environmental challenge that simply did not seem to exist?"

A: "Again, that question is an example of the fact that you simply do not understand science and how science works. Photosynthesis evolved by natural selection of random genetic mutations in answer to an environmental challenge to the survival of bacteria. We don't need to know the specifics of that challenge. The fact that the trait evolved is proof itself that the environmental challenge existed."

Q: "So, you assume that there just had to be an energy problem that needed to be solved in order for bacteria to develop the complicated process of photosynthesis?"

A: "Yes. That is a core assumption of current evolutionary theory."

Q: "What confirmable evidence do you have in support of that assumption?"

A: "That is the very foundation of the universally-accepted theory of evolution by natural selection. Let me read to you a section from our 1999 publication.

> 'As the great geneticist and evolutionist Theodosius Dobzhansky wrote in 1973, nothing in biology makes sense except in the light of evolution.'

Do you start to understand now?"

Q: "Frankly that doesn't seem like much of a scientific explanation to me. Yet, your scientific assumption is that an evolutionary challenge was presented to rock-eating bacteria and, even though we have no idea what the challenge was, the challenge was best met by those bacteria who evolved the ability to harvest the energy of the Sun as their alternative energy source?"

A: "Yes."

Q: "And, you are quite certain that the evolution of photosynthesis in bacteria was achieved by undirected random mutations of the genetic DNA of their ancestral rock-eating bacteria?"

A: "A crude way to put it, but yes."

Q: "I guess when the first bacteria evolved the process of photosynthesis that was certainly an example of an 'adaptive trait'. How did that single-celled organism develop the ability to do that?"

A: "Mr. Darrow this is becoming tiresome. We have covered this before, but I will cover it again. Photosynthesis is indeed an adaptive trait. It was developed through the natural selection of random mutations of genetic DNA. Again, the mutations were random, not directed. Mutations that proved to be useful to the survival of the species were retained in the species' gene pool. The series of mutations that resulted in developing the process of photosynthesis was most certainly useful for survival. Organisms that developed photosynthesis now derived the energy they needed for life directly from the energy of the Sun."

Q: "So, the development of the process of photosynthesis that you described was not in any way planned or directed. It evolved to meet an environmental challenge and it occurred by mistakes and accidents. Is that correct?"

A: "I would not put it in those words, but that is essentially correct."

Q: "Photosynthesis therefore occurs naturally. I recently read an article in a scientific journal about a team of scientists who are trying to develop photosynthesis in the laboratory. The term used was 'artificial photosynthesis'. So, to date scientists have not been able to create photosynthesis?"

A: "No, we have not. But we are working on it. Think of the reduction in fossil fuels and green house gases if we could just achieve that. And, I think that surely one day we will. That is the path of science."

Q: "Dr. Tall, it is most difficult for me, as a layman, to understand how random mutations of genetic DNA, without direction, could produce the elegant process of photosynthesis when the most brilliant scientific minds of today's scientists cannot figure out how to do it. How can that be?"

A: "It took nature ½ a billion years to evolve photosynthesis. We have been working on it for but a few years. The marvel of evolution by natural selection is that 'life will always find a way'. Are you starting to understand?"

Q: 'Professor Tall, to be honest, that is a pretty flippant answer. Nature's process as you describe it is mistake after mistake after mistake, without any direction or purpose. Your contention is that, without direction or purpose, nature can solve a very intractable problem but brilliant human minds that are very directed and very purposeful cannot. The bottom line question is this. What scientific proof is there that the process of photosynthesis evolved through nature's selection of **random** mutations of genetic DNA?"

A: "The proof, Mr. Darrow, is in the process of evolution by natural selection itself. Science only deals in natural processes. And, random mutation is the only explanation that comports with natural processes. Any other explanation belongs in the realm of metaphysics and religion. The theory of natural selection itself has been confirmed again and again by scientific evidence and is rock solid."

Q: "So, in essence, the mutations have to be random in order for the explanation to be scientific. Any directed or planned mutations would be unnatural and, by definition, outside of the scope of science. Is that it?"

A: "Mr. Darrow, you are not a scientist and you seem incapable of understanding the scientific method. If you want to believe that photosynthesis came about by some sort of miracle, that's fine with me. You can believe anything you want, but that is not science. The fact is that science deals only in natural explanations and only natural explanations should be presented in science class. Maybe you should study science a little more or simply defer to those of us who have far more knowledge of the natural world than you do."

Q: "Ouch. That does sting. But, frankly that answer provides absolutely zero scientific information. The Tennessee *'Only Science in Science Class Act'* that the NAS solidly supports provides this specific criteria:

'The teaching of science in science classes shall be limited to only natural explanations that can be inferred from confirmable data whose results can be observed, tested, replicated and verified.'

Unless you can provide specific confirmable data whose results have been observed, tested, replicated and verified to support the inference that photosynthesis is an adaptive trait that evolved by random genetic mutations then that explanation is indeed **not** scientific. Can you do that?"

A: "I already have. You simply cannot understand it."

Q: "I may not be a scientist, but I most certainly do understand the scientific method. The real power of the scientific method that has led to all the great discoveries of science is that scientific explanations must be based on confirmable empirical data and must always be subject to falsification. Again, what is the confirmable empirical data in support of your explanation and how can it possibly be falsified?"

A: "I have already given you my answers. I find no need to repeat myself yet again."

Q: "Well, let me summarize then with this question. Professor, which of the following explanations regarding the evolution of genetic mutations that resulted in the adaptive trait of photosynthesis is the more scientific to present to youngsters in public school science class?
- The mutations of genetic DNA were undirected and random.
- The mutations of genetic DNA were either undirected and random or they were directed and purposeful, and no one knows which is true."

A: "Sir, I know the answer to be a natural one. The genetic mutations were undirected and random. That is the only acceptable answer in science class."

Animals Un-Evolved Photosynthesis

Q: "So, your scientific explanation is that the adaptive trait of photosynthesis evolved over a period of ½ billion years through the process of natural selection of random mutations of genetic DNA.

After nature evolved the elegant process of photosynthesis in bacteria and green plants, why would the animals who evolved later in time not have retained the most-useful trait of being able to obtain the energy that they need to live directly from sunlight?"

A: "Again, Mr. Darrow, you are not a scientist and apparently cannot understand the science of evolution by natural selection. But, let me try again. Some primitive animal organisms, like some corals and sponges, did acquire photosynthesis through a symbiotic relationship with plant algae and cyanobacteria that provided photosynthetically-fixed carbon to the host animal. But the process of evolution did not continue photosynthesis in higher animal species."

Q: "Why didn't more complex animals like worms, and fish, and reptiles, and mammals retain the wondrous ability to receive the nourishment needed for life directly from sunlight?"

A: "By the time that animals of any size evolved, the photosynthesizing bacteria and plants had greatly changed the Earth's atmosphere. The ecosystem had evolved to include an oxygen rich atmosphere and living organisms were needed who would serve to balance things out in that new environment. So, the process of evolution by natural selection selected those random mutations that would result in the survival of living organisms who would derive the energy they needed by eating photosynthetic life. These new creatures thereby evolved a respiratory system whereby they extracted oxygen from the atmosphere and respired carbon dioxide back into the atmosphere where it was needed for photosynthesis by green plants and cyanobacteria."

Q: "Again, I'm sorry to be so dense. But, that does not explain why the useful and elegant process of photosynthesis would not have been retained by animals as they evolved. Why would they have lost that ability and instead be required to eat other living things or things that had once been alive?"

A: "I am also sorry that you seem to be incapable of understanding it. I have just explained it. I wish you could understand."

Q: "Okay. So, through the process of evolution by natural selection of random genetic mutations the scientific explanation is that as animals evolved they lost the ability to obtain energy directly from sunlight because losing that ability aided them in their survival.

    The loss of that ability, I guess, had to be an adaptive trait or else it would not have evolved. I must admit that it is difficult for a layman to understand how it would benefit an animal to lose the ability to create its own food supply. You are right. I am not capable of understanding that explanation because it just doesn't make sense."

* * *

At this point Clarence Darrow noticed Goody Spyer artfully engaged in a comb-over. He asked the Judge for a brief recess before proceeding with his questioning. Judge Raulston granted the request, and Clarence and Goody huddled together in conference.

Goody Spyer advised the attorney that he had about worn out the randomness issue in DNA concerning photosynthesis. And, he needed to stop belittling the esteemed scientist Noah Tall. Belittling usually produced a sympathetic ear for the person being belittled.

Goody advised Clarence that he needed to finish today's session by providing a wider scope to show the very incredulity of using randomness to explain the evolution of higher organisms. And, the segue he needed to get there from here in a coherent manner was to examine the phenomena of 'activation energy' and the use of enzymes. Clarence had made a big point of that in the briefing points of what he wanted to cover in that day's session but had failed to bring it up so far in the trial.

As the courtroom session reconvened Clarence Darrow took Goody's advice as he continued.

* * *

Animals Need Enzymes to Keep from Burning Up

Q: "Dr. Tall, before I continue with my questioning I would like to apologize. I got carried away with the subject of photosynthesis, I guess because it just seems so extraordinary. But, that is no excuse for being rude to you. I am sorry.

Well, enough about photosynthesis.

You have explained that after green plants perfected the process of photosynthesis, animals evolved who would no longer use that process to get the energy they need for living. Animals evolved to acquire energy by either eating the green plants or eating other animals that had eaten those green plants. Is that correct?"

A: "I wouldn't put it in those words, but that is essentially correct."

Q: "When animals like us eat plants, how do we extract that plant energy?"

A: "The energy is contained in the glucose sugar that the plants constructed through the process of photosynthesis. When we eat the plants we acquire the energy that is held in the chemical bonds that comprise the glucose sugar."

Q: "Could you give us the basic explanation for how we actually extract the energy from glucose sugar?"

A: "Mr. Darrow, the process is actually quite complex. But I will explain it as simply as I can.

In a living organism, metabolism is the sum of all of the catabolic and anabolic energy reactions. Catabolic reactions release energy and anabolic reactions add energy. Catabolic reactions provide the energy that drives the anabolic reactions forward.

The glucose sugar that we eat is a high-energy organic molecule. It is high in energy content and it is highly stable. Each cell of a living organism contains a chemical processing plant whereby the high-energy molecules are broken down in a step-by-step fashion to supply us with the energy we need. The cellular metabolic processes consist of a 10-step process called glycolysis and a process called cellular respiration. Cellular respiration is actually a system that is itself comprised of an 8-step process called the Krebs Cycle followed by the electron transport chain.

The electron transport chain then links to the process of oxidative phosporylation in order to produce the ubiquitous molecules of energy that we need for life. Those energy molecules are called ATP.

The total energy that is released and then transformed through this highly complex process is about 38 molecules of ATP for every molecule of glucose consumed. The amount of the initial energy that is stored in glucose that is recovered through our cellular chemical processing plant is about 40%.

Do you want me to explain each of these processes in more detail, or is that sufficient?"

Q: "Your explanation contains more than enough detail for us to begin to absorb.

So, the breakdown of glucose sugar is a catabolic process that runs downhill to provide us with useable energy. Does the breakdown of glucose automatically begin after we consume the glucose?"

A: "No. As I explained, the chemical breakdown begins after the organic molecules are delivered to the individual cells. Then the individual cell has to first destabilize the high-energy organic molecule in order for the catabolic process to begin. The chemical bonds have to be broken before the downhill process can begin. And, breaking the bonds requires an 'activation energy'. Activation energy is the amount of energy required to bring all the molecules in a chemical reaction to the reactive state."

Q: "Where does that activation energy come from?"

A: "The chemical activation energy comes from the thermal excitation of the high-energy organic molecules. Thermal excitation is just a description of the rate at which excited molecules bump into each other.

A chemist in the laboratory can supply the high level of activation energy required to get things started by exciting the molecules with heat. He can simply turn up the flame of the Bunsen burner in the lab. However, if that much heat were added within a living cell the cell would die. You can see the problem. We would spontaneously combust.

So, nature evolved the process of using specialized proteins, called enzymes, to supply the activation energy required to begin the catabolic process. Without enzymes acting as catalysts transition states are achieved only with the input of heat. Enzymes are great catalysts because they increase the reaction rate without increasing the temperature.

Enzymes speed the rate at which a chemical reaction occurs not by adding heat but, rather, by lowering the activation energy required to a level within the range of moderate temperatures within living cells.

That is how nature solved the problem. That is why we don't spontaneously combust as we gain the energy we need for living that is contained in high-energy organic molecules that we eat."

Q: "Dr. Tall, that is an incredible process. Let me make sure that I correctly understand. When we process the food that we eat we need an activation energy other than just adding heat in order to begin the catabolic process. Otherwise we would simply spontaneously combust. Enzymes evolved to solve that potentially fatal problem by increasing the chemical reaction rate without adding heat.

It seems to me that some sort of unknown, but built-in, problem-solving mechanism must have been used to develop such protein enzymes. If **random** mutations of genetic DNA was the mechanism that evolved such enzymes wouldn't the evolutionary result have quickly been the extinction of the first animal species?"

A: "Of course not. The change mechanism was in fact randomness. And, quite obviously all animal life did not die out."

Q: "Professor, let me explain why I find that answer to be unsupported by the scientific evidence.

You have told us that scientific evidence shows that enzymes are themselves proteins. And, like all proteins, they are constructed in accordance with the information instructions contained in the organism's DNA.

You explain that those enzyme proteins evolve only as copying errors occur in the organism's genetic DNA. Such copying errors occur very infrequently (less than one in a billion) and that when they do occur they very seldom (almost never) prove to be useful for the species. Therefore, when an adaptive change does occur it is invariably the result of a very, very long process. You tell us these improbabilities are accounted for because the changes occur over the course of evolutionary time (hard for us to even imagine).

Now, with enzymes, this is the problem I see with your explanations. The mutation that resulted in the adaptive change that produced enzymes had to be perfect the first time that a mutation occurred. Otherwise instant heat death by spontaneous combustion would have been the result when the first animal consumed a green plant. Is that not correct?"

A: "Yes. That is **not** correct. Adaptive change also can occur through random mutations that happen in a single generation, like we have seen in some bacterial mutants. Enzymes may have evolved through the same random mutation process that allows bacteria to quickly become resistant to antibiotics."

Q: "So, no matter how inconsistent it is to argue both sides of the coin, the answer is always randomness. How can that possibly be a scientific explanation? Isn't your answer simply based on your religious zeal?"

A: "Mr. Darrow, that is offensive. In our book *Science and Creationism* we state the explanation as clearly as we possibly can:

> 'It is the job of science to provide plausible natural explanations for natural phenomena.'

The explanation of randomness is the only possible plausible natural explanation for how life evolves through the process of evolution by natural selection. I just wish you could understand that. To be honest, your constant questioning of that scientific truth is getting quite tiresome."

Q: "You also explain in that same book that any creation that is not explained as a product of randomness cannot be included in science because that would indicate purposefulness. You explain that scientific creationism cannot be included in science class by this rationale:

> 'The arguments of creationists are not driven by evidence that can be observed in the natural world. Special creation or supernatural intervention is not subjectable to meaningful tests, which require predicting plausible results and then checking these results through observation and experimentation. Indeed, claims of "special creation" reverse the scientific process. The explanation is seen as unalterable, and evidence is sought only to support a particular conclusion by whatever means possible.'

Dr. Tall, we agree fully that creation science based on the explanation of purposefulness should be excluded from science class for the very reasons you provide in this excerpt. However, those same reasons apply exactly the same for your overarching explanation of randomness that you use to explain every aspect of biological development. The explanation of **randomness** is seen as unalterable, and evidence is sought only to support that particular conclusion. Is that not true?"

A: "Quite the contrary. That is not true. The reason for the randomness explanation, as I have repeated I don't know how many times, is that it is the only possible explanation if we are to remain scientific."

Q: "In the 4th century BC Aristotle explained that heavier objects accelerated faster than lighter objects.

   Everyone accepted that scientific 'truth' until Galileo showed by experiment that gravity accelerates all objects at the same rate. Isaac Newton then explained that gravity was a 'pull force' that was universal. Then Albert Einstein explained in his theory of Special Relativity that what we experience as the force of gravity is actually the result of the curvature of space when two objects near each other.

   Galileo, Newton, and Einstein were each possessed of scientific genius and they thought 'outside the box' of the scientific paradigms and beliefs of their times. As each step was taken in discovering a richer understanding of the mysterious force of gravity these icons of science were mocked by the scientific authorities of their times.

   The scientific elite of each era had faith in the scientific paradigm of their time and zealously adhered to their preconceived scientific 'truth'. Based on the faith of their religious zealotry they explained that their preconceived scientific 'truth' was, in essence, the only possible explanation if they were to remain scientific. They thus denounced each new scientific theory until confirmable scientific evidence became overwhelming.

   Professor Tall, isn't it possible that your insistence that **randomness just has to be the answer** smacks of that same kind of religious zeal?"

A: "Of course not. You have harped and harped that my answers are not scientific. I tell you that they are based on nothing less than a lifetime of scientific endeavor."

Q: "I know that I have harped on the point a lot and that it has become tiresome to you. I simply want to understand your explanations and to see if real scientific evidence exists in support of them. I'll try to use as little 'harping' as possible.

Let's proceed to learn a little more about the work of enzymes if you don't mind. Is there one particular enzyme that catalyzes the chemical process of converting glucose into ATP energy that we can use?"

A: "The specific enzyme that initially catalyzes the oxidation of glucose is called glucose oxidase. But, many different enzymes are necessary to perform the energy release and conversion process. For example, 10 different enzymes, each of which occurs in the cytoplasm of the individual cell, catalyze the 10 steps of glycolysis.

Enzymes are proteins that function as highly selective biological catalysts. They speed the rate at which a chemical reaction occurs by lowering the activation energy to a level in the range of normal cellular temperatures. The effect that enzymes have on the rate of chemical reactions is astronomical, effectively turning chemical reactions on and off.

There are over 2,000 known enzymes, and each enzyme typically catalyzes only a single reaction in only one kind of molecule. Each enzyme performs its function by tightly binding to a specific chemical molecule that is being changed. And, enzymes function at lightning speed, with a typical enzyme acting on about 1,000 molecules a second. You see, after the enzyme changes a molecule, the changed molecule drops the enzyme off so that it can bind to and change another molecule. The enzyme is never used up in this process.

We, of course, don't eat just glucose sugar. We eat all sorts of other carbohydrates and fats and proteins and nucleic acids. All of these things that we eat are high-energy organic molecules, and the same process is required in order to obtain from them the usable ATP energy that we need to live. And, all of these complex molecules require specific enzymes to carry out the catabolic energy process."

Q: "So, nature developed enzymes to provide the 'activation energy' that is needed to begin the catabolic energy process without us just burning up. And, the enzymes are themselves proteins that are constructed in each cell according to the coded information contained in DNA. Isn't that correct?"

A: "Yes. And, I can just imagine where you are going with this."

Q: "The NAS explanation for how each of those 2,000 enzymes came into existence was by a random mutation of the information in genetic DNA. Right?"

A: "Yes. That is correct."

Q: "We have stipulated in this trial that scientific explanations must be limited to those that can be inferred from confirmable data.

What is the specific confirmable data from which this randomness explanation can be inferred?"

A: "The randomness explanation for how each of these enzymes came into existence is based on the same confirmable data as that which supports the inference of randomness for how photosynthesis came into existence. Let me repeat. Science deals only with natural processes. And, random mutation is the only explanation that comports with natural processes."

Q: "Professor, simply saying something does not make it true. Scientific truth must be based on scientific evidence.

Let me ask you this final question about the evolution of enzymes which, among other things, allows us to gain the energy we need from plants without burning up. Which of the following explanations regarding the evolution of genetic mutations that resulted in specialized proteins that we call enzymes is the more scientific to present to youngsters in science class?

- The mutations of genetic DNA were undirected and random.
- The mutations of genetic DNA were either undirected and random or directed and purposeful, and no one knows which is true."

A: "You already know my answer, Mr. Darrow. The only acceptable answer in science class is that the mutations were undirected and random."

* * *

Judge Raulston called a recess. However, before he allowed the parties a bathroom break he again held a side bar conference with both attorneys. He asked William Jennings Bryan yet again if he wanted to continue to refrain from objections or cross-examination. Defense counsel repeated his stance and the Judge ended the side bar.

Clarence Darrow then had the opportunity to again confer with Goody Spyer who told him he was now doing a better job of making his points. He needed now to end the day's session by presenting a 'big picture' summary.

The Judge reconvened the courtroom proceedings and Clarence Darrow continued with his questions.

* * *

From Single-Celled Organisms to First Animals

Q: "Professor Tall, would you now please give us an overview thumbnail explanation of how life evolved from the first single-celled bacteria to the first animals?"

A: "Certainly. When the first life appeared on Earth about 3 and ½ billion years ago it took the form of single-celled organisms without a nucleus. Over the next 2 billion years those single-celled organisms evolved a nucleus to house DNA within a protective membrane.

These single-celled organisms evolved other separate internal cell structures called organelles, which are each surrounded by their own additional membrane. These organelles evolved to perform specialized functions. For example an organelle called mitochondria synthesizes useful energy for both plants and animals. Organelles called ribosomes translate the information contained in nucleic acids into the language of amino acids necessary to construct the proteins needed to perform all the functions for life and living. Organelles called lysosomes evolved to rid the cell of the waste products of cellular metabolism. We are quite confident that these organelles evolved by a process called endosymbiosis whereby one single-celled organism was consumed by another and the consumee became a functional part of the consumor. Then the stage was set for multi-celled organisms to evolve.

Thereby, over a period of almost 3 billion years very simple multi-celled organisms evolved. The first multi-celled organisms were invertebrate (had no backbone) and seem to have evolved through an aggregation of single-celled organisms, called flagellates. Scientists believe that the origin of virtually all the members of both the plant kingdom and the animal kingdom may be traced to such flagellates assembled into large groups. By this manner, cells comprising both plants and animals developed an interdependence and an elemental form of symbiosis (mutual dependence).

So, after about 3 billion years of development, life had become 'advanced' in terms of DNA, nucleated cells, and multiple-cell organism beginnings. But, the nature of such life still remained very small, co-existed with the much more numerous bacteria, and remained, itself, in very simple forms. Up to about 540 million years ago the most advanced life on Earth still lacked distinctive features, had no mouth or anus, no head or tail, and was not capable of self-locomotion. All life still remained in the seas, and the most advanced form of life, the jellyfish, simply drifted along at the whim of ocean currents.

And then about 540 million years ago complex life burst forth on the planet in great profusion and diversity in a very short period of time (in terms of geological time). That began what is termed the Cambrian Period explosion.

With protection from harmful ultraviolet radiation, with a fairly healthy atmosphere, and with quite a warm climate, little organisms of complex life began to proliferate not only in the ocean but also in shallow pools of water inland.

Complex animal life began with a metaphorical bang at the start of the Cambrian Period about 540 million years ago. If you look at the 3 and ½ billion years of development from the beginning of the first life until the present as a 24-hour clock of geological time, the profusion of life that appeared over the thirty million years comprising that period took place in about 15 minutes of life's time line.

Evolution did not proceed along a straight line leading from bacteria to human beings. The evolutionary process was very much non-linear, with life developing in different forms and variations much like thorns and twigs and branches of an ever-growing and very thorny bush.

The first multi-cellular sponges had no nerve cell. Next, the jellyfish family developed two layers of cells, and, importantly, bilateral symmetry (a left and a right side and a head and a tail end) and locomotion. Flatworms developed three layers of cells (basic primary tissues) and a complex nervous system, and the worm-like animals that followed (planarians) had it all: bilateral symmetry; a mouth and anus; a complex central nervous system; locomotion; and a defined body cavity. The increase in morphological complexity was immense over this brief period. And the variety of different bodily plans developed during this period is believed by some scientists to be greater than those living in the world's oceans today. Indeed, in all the time from the Cambrian Period to the present, very few new basic body plans have been added to the ones that apparently originated in the Cambrian Period (540-510 million years ago). And, all of that is very strong evidence in support of evolution by natural selection."

Q: "Wow. That is amazing."

A: "You asked for a thumbnail explanation and that is pretty much it."

Q: "An early-on adaptation that you mentioned was symbiosis, a mutual dependence between living organisms. Did that mutual dependence develop through the mechanism of random genetic mutations?"

A: "Of course it did."

Q: "As I understand it, such mutualistic partnerships, or symbioses, actually began with single-celled organisms that had yet to develop a nucleus, the so-called prokaryotes. So, this mutual partnership between living organisms would have developed quite early and has been retained by many species ever since. Is that right?"

A: "That is correct."

Q: "Such mutualistic partnerships are indeed quite commonplace in our world today. The most obvious example is that of the birds and the bees and the flowers. Permit me to read a passage from a primer on flower pollination as a predicate to my next question.

'As a flower blossoms, the beauty that unfolds serves as an attractant to the literal "birds and bees". Invitation is extended to visitors from the animal kingdom. Different colors attract different visitors. Bees are attracted to blues and hummingbirds are attracted to red, for example. And, other birds, and insects, and butterflies, and bats, and wasps, and flies, are attracted by other colors and scents and flower shapes. Because of their unique shape, some flowers are only able to be pollinated by a specific species of insect or bird.

The animal visitor is an unwitting participant in the plant reproductive process. What the animal visitor is in fact attracted to is food. The food is in the form of a very nutritious liquid called **nectar**. The nectar is produced and reposes in special glands, called **nectaries**, which are located at the base of the pistils and stamens. As the animal visitor attempts to reach the nectar reposing in the nectaries, the pollen is transferred to him. The hungry visitor from the animal kingdom can't help but get the pollen on him. He literally cannot miss. A flower's contrasting colors seem to serve as a **nectar guide** actually directing the visitor to the pollen source.

As the avian or insect visitor reaches the nectar source, as it hovers and feeds, the pollen is transferred to its body. As it then visits a second plant in further search of nectar, it thereby inadvertently transfers the pollen onto the sticky stigma of the second plant. Pollination starts at that point and ends with the fertilization of the ovule reposing in the plant's ovary.'

Now, let me ask this. How is it possible for unrelated species to each modify their genetic DNA by the mechanism of randomness and end up with a mutual benefit? Haven't we now reached the point of absolute absurdity in contending that the adaptive trait that evolved for each species was the result of natural selection of random mutations? Such symbiotic relationships simply cannot occur randomly, isn't that so?"

A: "Of course they occur by natural selection of random mutations. This one is, in fact, quite easy to explain. Even you should be able to understand Mr. Darrow.

The fossil record provides solid evidence that the earliest seed-bearing plants were non-flowering plants. Their pollen was simply dispersed by the wind. As winged insects evolved they began to eat the highly nutritious pollen. That was detrimental to the plant. So, nature selected mutations that made the pollen more likely to stick to insects who would, in turn, provide a more effective pollination mechanism than the wind. More and more random mutations were selected to produce colorful flowers that would attract the insects and that would make the pollen even more adapted to sticking to the insects. Thereby the initial harm that was caused by the insects was equalized by the benefit.

I do hope that you can understand this very clear example of the evolution of mutualistic partnerships by the natural selection of random mutations of genetic DNA."

Q: "Professor Tall, there is absolutely nothing about that explanation that provides any confirmable evidence whatsoever that **random** mutations produced the quite wondrous result of mutual benefit.

I guess that you simply expect all of us to accept your explanation on faith because you are a world-renowned scientific authority. Your credentials are impeccable, but those impeccable credentials do not constitute confirmable scientific evidence.

Well, I'll just resolve myself to frustration at this point and move on. Professor Tall, that takes us through about 3 billion years of evolution by natural selection of adaptive traits. And, just so I am clear, your scientific explanation is that all of those adaptive traits evolved by nature selecting mutations that were useful for the survival of the species. And the mutations were random and undirected and without purpose. Is that correct?"

A: "Mr. Darrow, no matter how many times you ask it, yes. That is correct. That is the scientific explanation."

Q: "Dr. Tall, let me recap just some of those adaptive traits:
   1. Information contained in DNA.
   2. Amino acids chaining together in the precise order necessary to form the discrete proteins that are needed in order for an independent living organism to function.
   3. Development of photosynthesis.

4. Development of enzymes.
5. Development of organelles within cells to provide for metabolism and waste removal.
6. Development of cellular layers for basic primary tissues.
7. Development of bilateral symmetry, with a left and a right side and a head and a tail end.
8. Development of locomotion.
9. Development of mutual partnerships (symbioses).
10. Development of a complex nervous system.

Now, let me ask this very important question again as it pertains to each of these adaptive traits. Professor, for the development of each of these adaptive traits, which of the following is the best scientific explanation to present to youngsters in public school science class?
• They evolved by the natural selection of DNA mutations that were undirected and random.
• They evolved by the natural selection of DNA mutations that were either undirected and random or that were directed and purposeful, and no one knows which is true."

A: "Again, Mr. Darrow, I know the answer to be a natural one. They were undirected and random. That is the only acceptable answer in science class."

Q: "In the 2nd century AD the Greek physiologist, Galen, explained that food traveled to the liver where it was transformed into blood. Veins from the liver carried blood to all the tissues and organs of the body. Blood was not pumped by the heart. Everyone accepted that scientific 'truth' until William Harvey, an English physician, published his explanation that blood was pumped throughout the body, from and to the heart. He explained that the blood supply of the body was recirculated again and again and that food was not transformed into blood.
Harvey's book was published in 1628. He was not the first to explain the blood circulatory system. In 1553 Michael Servetus, a Spanish physician, published a book describing

the pulmonary transit of blood.   His explanation resulted in Servetus being burnt at the stake.   Harvey still had great trepidation about his book that challenged the scientific paradigm and belief of the times. He feared for his life.

For many years the scientific elite of the times disparaged Harvey's theory.   Some contended that the circulation of the blood would deliver 'putrid' material to the entire body. Others contended that circulation would 're-cook' the blood and turn it into bile. Over 20 years after Harvey published his blood circulation explanation, one of the most elite scientists of the times, Jean Riolan, dean of the Medical Faculty in Paris, still explained that blood was produced in the liver and that pulmonary circulation simply did not exist.

The scientific elite of each era had faith in the scientific paradigm of their time and zealously adhered to their preconceived scientific 'truth'.   Based on the faith of their religious zealotry they explained that their preconceived scientific 'truth' was, in essence, the only possible explanation if they were to remain scientific.   They thus denounced each new scientific theory until confirmable scientific evidence became overwhelming.

Professor Tall, isn't it possible that your insistence that **randomness just has to be the answer** smacks of that same kind of religious zeal?"

A: "Of course not. My answers are based on rock solid science. They are based on nothing less than a lifetime of scientific endeavor."

Q: "Professor Tall, we have stipulated that you are the most eminent scientist who is most qualified to provide scientific testimony in this trial. But, who are you really representing here today, science or Darwin's God?"

* * *

With that rhetorical question Clarence Darrow 'accidentally' knocked his notes from the lectern onto the floor and addressed Judge Raulston.

"Your Honor, the information in my notes just had an accident. I request a recess so that I may reorder the information contained in my notes into a sequence that proves to be helpful, not harmful."

The Judge reproved Clarence Darrow for employing a most-amateurish trial tactic. Then he called a brief recess. After again conferring with Goody Spyer, Clarence Darrow resumed his questioning.

Phenotypes, Genotypes and Cancer

Q: "Professor Tall, a living organism, like you or me, has both a phenotype and a genotype. The phenotype is the actual form and function of our body as we go about our job of living in the natural world. The genotype is the information in our DNA that provides the instructions necessary to build and operate and maintain the phenotype. Is that basically correct?"

A: "That's a crude description, but it is essentially correct."

Q: "And, modern science has discovered overwhelming evidence that changes in the phenotype through evolution can only occur through changes in the genotype, in our DNA. Is that basically correct?"

A: "Yes."

Q: "And, evolution by natural selection tells us that nature selects changes in the phenotype that prove useful to the survival of the species. Those beneficial changes are called 'adaptive traits'. And, the mechanism from which nature selects beneficial changes in the phenotype is random mutation of the DNA genotype. In essence, the NAS explanation is that random mutation in our DNA produces beneficial change through evolutionary time. Is that correct?"

A: "Crudely put, but, yes, that is correct."

Q: "All adaptive change is thereby explained as the result of
    random mutation that occurs in one of two ways:
    • A mistake during DNA replication in a sex cell, or
    • Damage to the DNA in a sex cell caused by some radioactivity
      or cosmic rays or some poison in the environment.

    Is that correct?"

A: "Yes, again."

Q: "Therefore, such wondrous adaptations as the development of
    the senses of hearing and smelling and tasting and feeling and
    seeing are, in fact, the result of genetic DNA molecules simply
    being misaligned or damaged. Still correct?"

A: "Mr. Darrow, you may try to paint a warped picture, but that is
    still basically correct."

Q: "Professor, I am not intending to paint a warped picture at all.
    In fact I am trying to paint a very clear picture. For the other
    side of that DNA coin paints a completely opposite picture.
    And, that presents a real conundrum.
        Is it not true that science has discovered that many deadly
    cancers are caused by damage to the DNA in our body's
    cells?"

A: "There are many causes and we are working on many cures for
    cancers. Yet, it is true that damage to DNA in our body's cells
    results in deadly cancers."

Q: "In essence the conclusion that can be drawn from this is that
    damage to our DNA over evolutionary time is good for us, but
    damage to our DNA during our lifetime will kill us.
        DNA mutations cause us pain and suffering during our
    lifetimes. DNA mutations during our lifetimes always cause us
    harm. They never improve anything.

Yet, DNA mutations in the long run provide exquisite improvements, like the development of our five senses and our elaborate circulatory and immune systems and wondrous organs, like the human eye and human heart and human brain.

Dr. Tall, does that really make sense to you?"

A: "Mr. Darrow, you do try to twist things around."

Q: "No, Professor Tall, I do not. I would simply like to know what the confirmable scientific evidence is to support the NAS explanation that both living harm and evolutionary benefit are provided by the same mechanism of randomness. Could you tell me that?"

A: "I have told you repeatedly. Randomness is the only possible explanation if we are going to remain scientific. You simply seem to be incapable of understanding that scientific truth."

Q: "In the 4th century BC Aristotle provided an elaborate scientific explanation that placed the Earth at the center of the universe. In 1543 Nicolaus Copernicus published a book that directly contradicted Aristotle and placed the Sun at the center of our solar system. For fear of retribution he waited to publish his explanation until he literally lay on his deathbed. In the next century Galileo Galilei observed through the newly-invented telescope astronomical phenomena that provided confirmable scientific evidence in support of the Copernican explanation. The elite scientists of Galileo's era devoutly believed in an Earth-centered universe that was contrary to his evidence. Based on their scientific authority, Galileo was tried, convicted, and put under house arrest until he died.

The scientific elite of each era had faith in the scientific paradigm of their time and zealously adhered to their preconceived scientific 'truth'. Based on the faith of their religious zealotry they explained that their preconceived scientific 'truth' was, in essence, the only possible explanation if they were to remain scientific. They thus denounced each new scientific theory until confirmable scientific evidence became overwhelming.

Professor Tall, isn't it possible that your insistence that **randomness just has to be the answer** smacks of that same kind of religious zeal?"

A: "Of course not. My answers are based on rock solid science. They are based on nothing less than a lifetime of scientific endeavor."

Q: "It is self-evident that the integrated complexity that has evolved over time for each living organism is enormous, and that certainly includes us *Homo sapiens*. As we discuss these things today, each of us has within our bodies an incredibly complex machine. We each have a blood works complex of some 50,000 miles of arteries and arterioles and veins and capillaries. And we have an energy processing plant that finely orchestrates the coordinated operations of our brain, heart, lungs, liver, kidneys, pancreas, gall bladder, spleen, stomach and intestinal system, to name but a few of our evolved organs and processes.

As we breath-in the air we need for life and digest our breakfast we give no thought whatsoever to those wondrous internal organs and functions that you tell us evolved by the natural selection of mistakes. That is, of course, unless something goes wrong. If mistakes occur in those wondrous internal organs and functions during our lifetime we experience pain, sickness or even death.

Is that a correct assessment, Professor Tall?"

A: "Yes."

Q: "Again, doesn't that seem strange and contradictory to you?"

A: "No. It does not. It is quite simply the 'scientific truth'."

Q: "The NAS explanation for how all of these organs and processes occurred, as I understand it is this. Nature, at some time in the past, simply selected from random mistakes in DNA to develop and coordinate the functions of all of these finely-tuned internal operations. Is that basically the correct explanation for how we evolved without the need of any intention or direction?"

A: "Yes, again. That is correct. There was no need for intention or direction."

Q: "Is there, in fact, any actual confirmable scientific evidence to support that explanation?"

A: "Of course there is. Otherwise we would not be teaching that explanation in science class."

Q: "If you would like to provide that specific evidence today, now would be a good time. Would you like to do that?"

A: "I have provided the specific evidence repeatedly, Mr. Darrow, and I will continue to do so. Perhaps as we move on to further matters you will begin to be able to actually see and understand that scientific evidence. I appreciate that such may prove difficult for a non-scientist."

Q: "In the 4th century BC Aristotle explained that the universe was static, unchanging, immutable, forever-existing. The Earth and the heavenly bodies had always existed. There was no creation event. Everyone accepted that scientific 'truth' until well after the beginning of the 20th century AD.

When Albert Einstein published his theory of General Relativity in 1916 it contained the most-disconcerting discovery that the universe must be either contracting or expanding. Einstein was so wedded to belief in the 'truth' of a static universe that he included a 'cosmological constant' that would serve as a kind of anti-gravity force needed to maintain a static universe.

In 1929 Edwin Hubble discovered through his telescope at Mount Wilson near Los Angeles that the stars in distant galaxies are moving away from us. His discovery was profound because it provided confirmable scientific evidence that the universe was not static. It provided support for the 'Big Bang' theory that explains that the universe has existed for about 14 billion years. We live in a universe of finite age and one that is expanding at an accelerating rate.

The scientific elite of each era had faith in the scientific paradigm of their time and zealously adhered to their preconceived scientific 'truth'. Based on the faith of their religious zealotry they explained that their preconceived scientific 'truth' was, in essence, the only possible explanation if they were to remain scientific. They thus denounced each new scientific theory until confirmable scientific evidence became overwhelming. After Hubble provided confirmable scientific evidence, a humble Albert Einstein declared his 'cosmological constant' to be his greatest scientific mistake.

Professor Tall, isn't it possible that your insistence that **randomness just has to be the answer** smacks of that same kind of religious zeal?"

A: "Of course not. My answers are based on rock solid science. They are based on nothing less than a lifetime of scientific endeavor."

* * *

With that answer, Clarence Darrow turned to address Judge Raulston directly. He indicated that he had no further questions for this day. Judge Raulston adjourned the proceedings till the following day.

CHAPTER 7

The Trial – The Fourth Day
Evolution of Electricity and Other Wonders in Animals

The next day in court began with another side bar discussion between the two lawyers and Judge Raulston. The Judge offered the Defense counsel yet another opportunity to begin to raise objections and engage in cross- examination. William Jennings Bryan remained adamant in his refusal. He indicated that he had discussed that very matter with Noah Tall the previous evening and the eminent scientist insisted that he did not want his lawyer's assistance in his dialogue with Clarence Darrow.

Upon the arrival of the two attorneys at their respective positions in the courtroom, the questioning of the fourth day of trial began. Clarence Darrow spoke.

* * *

Electrical Circuits in Animals

Q: "Dr. Tall, you have been quite patient with my questioning, and I thank you for that. It is not easy for me to ask pertinent questions relevant to the outcome of this case without exploring some very significant evolutionary developments. When we adjourned yesterday, you had patiently explained the basics of evolution from the first life on Earth up to and including the evolution of the first multi-cellular animals like worms and sponges and jellyfish.

With the evolution of that first primitive animal life wasn't there a very significant evolutionary development involving electricity?"

A: "Yes. As animals evolved they developed nerve nets composed of neurons. Neurons are specialized cells that transmit information by electrical current."

Q: "What were the first organisms to evolve the adaptive trait of transmitting information by electrical current?"

A: "The tiny invertebrate animals, those without a backbone, evolved first. Even primitive sponges with only one cellular level of organization evolved the ability to respond to stimuli by the coordination of electrical charge. Next, the hydra developed two cell layers and a nerve net composed of neurons in contact with one another. Then the jellyfish evolved two nerve nets: the fast acting nerve net enabled major responses to environmental dangers while another nerve net coordinated slower and more delicate movements."

Q: "As I recall your testimony, the jellyfish evolved shortly before the beginning of the Cambrian Period, sometime before 540 million years ago. Is that correct?"

A: "Yes."

Q: "What neural development occurred over the 30 million year period that comprised the Cambrian Period?"

A: "An absolutely amazing amount of neural development occurred in that period. Following the jellyfish, the nervous system of the planarians developed bilateral symmetry with two lateral nerve cords that allowed for rapid transmission of information from anterior to posterior. They were the first to develop a simple brain composed of a cluster of neurons, called ganglia. The cerebral ganglia receive input from photoreceptors in eyespots and sensory cells. The transverse nerve fibers between the sides of the ladder-like nerve cords keep the movement on both sides of a planarian body coordinated. This planarian nervous system foreshadows the central and peripheral nervous systems that would evolve later in vertebrates. By the way, that provides an excellent example of evolution by natural selection.

Then the arthropods, like crustaceans, spiders, and insects evolved as invertebrate animals with true nervous systems. They have a brain that receives and processes sensory information and controls the activity of the ganglia so that animal functions are coordinated.

And then the first vertebrate animals evolved toward the end of the Cambrian, about 510 million years ago. The fossil record is clear about that. The first vertebrates appeared as fish in the World Ocean. They had both a central and a peripheral nervous system. The central nervous system develops a brain and spinal cord from an embryonic dorsal nerve cord. The peripheral nervous system consists of paired cranial and spinal nerves."

Q: "As I understand it, each neuron is a transmitter of information. How many neurons are we talking about that serve to electrically transmit information in the simple invertebrate animals?"

A: "An insect may have as many as one million neurons. Multiply that by about a hundred thousand times for vertebrate animals like us."

Q: "I am almost afraid to ask, but I will. How do neurons function in order to transmit information? What is the simple explanation?"

A: "The simplest explanation is that neurons function by changing the permeability of sodium and potassium ions across cell membranes. That is basically how an electric charge is generated for transmission to the next neuron adjacent to it. It is pretty complicated, but that is it in a nutshell."

Q: "Are neurons the only cells in a living organism that generate and maintain an electric charge?"

A: "No. Far from it. Every animal that has ever lived on Earth that is the size of an insect or larger, including us humans, requires electrical charge to be maintained within its cellular structure. If that electrical charge is not maintained, death results."

Active Pumps are Essential for Animal Life

Q: "How is the electrical charge within animal cells maintained?"

A: "By what we call sodium-potassium pumps. Those pumps are employed to maintain a constant electrical circuit."

Q: "So, electrical charge would not be maintained without the operation of the sodium-potassium pumps in the cells of animals?"

A: "That is correct."

Q: "Evolution of an actual pump seems pretty advanced. At what stage of animal development did the adaptive trait of the sodium-potassium pump evolve?"

A: "They evolved quite early. In fact, the sodium-potassium pumps are actually found in some bacterial life which were antecedent to animal life. You see, Mr. Darrow, this is again quite a compelling piece of scientific evidence for the proof of evolution by the natural selection of random genetic mutations producing quite adaptive traits. The adaptive trait of sodium-potassium pumps which are necessary to maintain electric charge is a prime example of what we scientists term a 'highly conserved' trait."

Q: "Could you please provide a thumbnail explanation of how a living cell maintains electric charge and how the sodium-potassium pumps operate to keep the cell electrified?"

A: "I'll try. I just hope that you will be able to understand this excellent example of evolution by natural selection of random mutations of genetic DNA.

Every living animal organism contains complex electrical circuitry that is used to communicate not only between cells, but also between the interior of the cell and the environment outside of the cell's plasma membrane. All living animal cells run on electricity and the electricity is generated by organic chemistry.

Maintaining cellular charge is a basic requirement for animal life. The failure to maintain proper electric charge in a living organism will result in organ failure and ultimately death. And, cellular charge must be maintained proactively.

Cells are constructed so that charged particles cannot pass directly through the plasma cell membrane. This feature lays the foundation for the electrical interactions. The inside of a living cell is constructed in such a fashion as to be polarized (have a negative charge). Basically, this is done by maintaining a high concentration of potassium ions within the cell compared to the outside of the cell, and a high concentration of sodium ions outside the cell compared to the inside. These ions move through specialized proteins that transverse through the cell membrane allowing for flow back and forth. They are known as ion channels. All of the ions naturally flow through passive ion channels in accord with the laws of physics from the place of highest concentration to the place of lowest concentration, as they strive to equalize concentrations. However, if concentrations become equalized there will be no voltage across the cell's plasma membrane."

* * *

Noah Tall paused in his testimony long enough to take a breath and a drink of water.

* * *

"A cell's membrane potential is the electrical potential difference (known as voltage) across the cell's plasma membrane. Voltage is caused by the separation of charges by the membrane.

It is stored energy that can be used to do work. Importantly, if ion concentrations become equalized there will be no voltage across the cell's plasma membrane. So, something else is required in order to maintain the cellular polarity that is required for a multitude of wondrous functions to occur within a living organism that are necessary for life. That something else is the cell's construction of **active** ion channels transversing the cell's plasma membrane, called sodium-potassium **pumps**.

Each cell constructs literally thousands of these sodium-potassium pumps to actively transport sodium and potassium ions back and forth between the cell's interior and the exterior of the cell's membrane. For each cycle of the pump, precisely three sodium ions are pumped out and two potassium ions are pumped in. All the pumps act the same way. Each time a pump operates, three sodium ions are pumped out and two potassium ions are pumped into the cell. The rate at which the pump operates is variable. The rate of pumping is regulated by hormones.

The sodium-potassium pump is an actual pump. It pumps the ions 'uphill', against their concentration gradients. And, that requires a lot of energy. It takes about one-third of our body's total resting energy to run these pumps. But, it is energy well spent. Without these pumps all living animals would cease to exist. Animal life itself depends on the maintenance of cellular polarity and our ongoing electrical circuitry. And, the sodium-potassium pumps are essential to the maintenance of cellular polarity."

* * *

Noah Tall concluded his explanation in this manner that, frankly, surprised Clarence Darrow. The Plaintiff's attorney remained silent as he referred to his notes for what seemed to be a full minute or so before continuing with his questions.

* * *

Q: "Professor, let me recap what you have just said so that I may clearly understand. In essence, what you are saying is that random mutations of genetic DNA produced a mechanism, an actual pump, that is used to reverse a natural process. Potassium ions naturally flow from a point of high concentration inside of the cell to a point of low concentration outside. Sodium ions do the opposite. Both flow in accord with the laws of physics. But, the end result of this natural process would be for all ion concentrations to equalize, which would result in no electric charge in the cell. Now, that did not happen. Rather, living cells figured out how to reverse the natural process in order to maintain electric charge. These ion pumps, that were developed early on in the evolution of animals, actually reverse a natural process in order to maintain an unnatural process — electric charge. How could **random** mutations of genetic DNA possibly produce a pump mechanism that reverses a natural process?"

A: "Quite simply. The animals that benefited from that random mutation were better suited to face environmental challenges. The sodium-potassium pump adaptive trait was retained in the gene pool of the species. That is quite obvious by the evidence that the pumps are a highly conserved trait, as I have already explained."

Q: "Again, that answer simply sidesteps the very issue at bar, whether genetic DNA mutations are random and undirected or not. What confirmable scientific evidence actually exists to support the explanation that the mutations of genetic DNA that resulted in the development of sodium-potassium pumps were in fact random and undirected?"

A: "I have already told you many, many times. It seems that you are simply incapable of understanding."

Q: "Dr. Tall, an authoritative pronouncement by an eminent and distinguished scientist, which you certainly are, does not amount to actual scientific evidence. All of us have agreed to a number of stipulations in this trial. A very pertinent one to the issue we are discussing is this stipulation, and I quote:

> 'In science, explanations are restricted to those that can be inferred from the confirmable data. Scientific explanations are bound by that limitation.'

Bearing that stipulation clearly in mind let me pose this question yet again. What confirmable data exists to support the explanation that a physical mechanism, an actual pump that reverses a natural process in order to maintain an unnatural one – electric charge – evolved by the process of random mutation of genetic DNA?"

A: "Mr. Darrow, again, I have already provided the evidence. The early occurrence of sodium-potassium pumps in some bacteria and early animals has become a highly conserved trait. That is scientific evidence."

Q: "No, Dr. Tall that is not scientific evidence. An after-the-fact explanation that these pumps evolved because they first occurred early-on in the development of animal life is based on no confirmable evidence whatsoever.

That kind of explanation is reminiscent of Aristotle's explanation that the Sun and the Moon and all of the other heavenly bodies were perfectly circular spheres. The 'scientific evidence' that Aristotle offered was that the heavenly bodies **just had to** be perfect circles because the circle is the shape of perfection. What shape could the purposeful gods possibly employ other than a perfect shape?

It seems to me that Aristotle's reasoning is simply the opposite of your reasoning that DNA mutations **just had to** be random because you believe that there is no Purposeful God. So your 'scientific explanation' boils down to this. What process other than randomness could possibly be employed by your Accidental God? Isn't that really what you are saying Dr. Tall?"

A: "Of course not.   I have already answered your question regarding confirmable scientific evidence. And, I really do not appreciate your sarcasm and innuendo."

Q: "I hear your answer, but I certainly don't hear a scientific answer based on confirmable scientific evidence. Professor Tall, let me ask this pointed question and please carefully consider your answer.

Which of the following explanations regarding the evolution by natural selection of genetic DNA mutations, that produced the adaptive trait of sodium- potassium pumps in all animal cells, is the more scientific to present to youngsters in public school science class?
- They were undirected and random.
- They were either undirected and random or they were directed and purposeful, and no one knows which is true."

A: "I've answered the question before, but I will be happy to repeat. They were undirected and random. That is the only acceptable answer in science class."

Q: "Well I thought that by this point you might have reconsidered, but obviously you have not."

* * *

Goody Spyer's comb-over was finally noticed by Clarence Darrow. Judge Raulston agreed to a fifteen minute recess during which the Plaintiff's counsel and his observant court watcher could confer.

All the talk about these pumps was getting quite overdone. The larger picture was being lost in the details. Goody advised that Clarence Darrow needed to get past these pumps and look at the marvel of how neurons passed on information through a combination of electricity and chemistry.

* * *

Q: "Professor Tall, what are some of the purposes served by the maintenance of electrical charge in animals?"

A: "Simply stated, our bodily systems can't work, our bodily organs like our heart and kidneys can't work, and our brain can't work without the operation of these sodium-potassium pumps. They maintain our internal cellular polarity and thereby provide the electrical circuitry that allows life to work.

The sodium-potassium pump that is used to maintain cellular polarity allows for many wondrous functions to occur within the body. To name but a few:
   • The export of sodium from the cell provides the driving force for many facilitated transporters, which import glucose, amino acids and other nutrients into the cell.
   • The export of sodium and import of potassium creates an osmotic gradient that drives absorption of water, which allows our small intestine and kidneys to work.
   • The rapid re-establishment of the resting potential of a neuron cell after it 'fires' enables a neuron to quickly 'fire' again and again and again in response to additional stimuli."

How Neurons Function To Transmit Information

Q: "The last point you made was about neurons 'firing'. You touched upon the development of neurons in your earlier testimony. As I recall, the evolution of the first nerve nets of neurons touching each other to transmit information occurred as early as the jellyfish. Could you elaborate on how neurons function – how they transmit information through electricity?"

A: "Sure. To keep it simple I'll use our own central nervous system as the example. But, bear in mind that all vertebrate animals have a central nervous system.

The central nervous system in mammals serves to coordinate instinctual behavior, body orientation, muscle coordination, and learning. The system resides in the brain and spinal column and is composed of a network of nerve cells called neurons.

The spinal cord is the portion of the central nervous system that is involved with reflexes and conducting nerve impulses from the peripheral nervous system (PNS) to and from the brain through 31 root pairs of spinal nerves. Each spinal nerve is attached to the spinal cord by a root of sensory fibers and a root of motor fibers.

The adult human brain is our on-board computer that weighs about three pounds. To run the computer requires about 20% of our body's total energy production. Our adult brain and nervous system contain about 100 billion specialized nerve cells, called neurons. Most of the brain is contained within our skull, but it is integrated into our spinal cord to provide us with a complete central nervous system (CNS). The CNS receives messages from the nerves that are in the tissues of the peripheral nervous system (PNS) and sends messages to the motor receptors of the PNS.

Neurons are cells that transmit electrical nerve impulses and respond to stimuli. Neurons look like and act like specialized cellular wires, much like electrical wires. Neurons run from the body of the cell located in the central nervous system to each and every part of the body through long processes (called axons) extending away from the body of the cell. So, the neurons that run from my central nervous system to my toes are several feet in length. At the end of each axon is a synaptic terminal that allows for the transmission of a signal to another cell, and so on and so on.

The messages to and from neurons are transmitted through an elaborate electrical circuit.

Motor neurons conduct electrical nerve impulses away from the brain or spinal cord. Sensory neurons conduct electrical nerve impulses from sensory receptor cells to the brain or spinal cord. And these electrical nerve impulses transmit information.

Neuron cells come in a wide array of shapes and sizes. Most are quite small, but some (like motor neurons that stretch from the base of our spine to the tip of our toes, and sensory neurons that stretch from the tip of our toes to the base of our spine) can be over three feet long. And neurons run to and from every part of the body, from just under the skin to the inner organs of the body, like the heart and lungs.

In many respects neurons are very much like other body cells. They contain cellular membranes, cytoplasm, organelles, and a nucleus housing the DNA library. However, neurons have certain characteristics that make them special. And, these special characteristics largely involve electricity.
The neuron nerve cells consist of four major parts:
- **soma** - the central part of the neuron which contains the nucleus of the cell;
- **dendrites** - the multiple input branches of a neuron that attach to the soma;
- **axon** - the long wire-like extension of the cell that carries electrical signals away from the soma; and
- **axon terminal** - the cellular output structure at the end of the axon that is used to release neurotransmitter chemicals and thereby communicate with other neurons.

At core, the story of neurons is an 'electrifying' tale. For neurons are electrically excitable cells in the nervous system that serve to process and transmit information.

Like all living cells in the human body, nerve cells (neurons) have **ions** (molecules that have either gained or lost an electron and thereby become **electrically charged**) in the fluids both inside the cell and outside the cell. And, like all living cells, there are **ion channels** transversing the cell's plasma membrane through which ion passage is accomplished. Ion channels are pore-forming proteins that control the small voltage gradient that exists across the plasma membrane of all living cells.

A cell is said to be 'at rest' when it is not stimulated. When a neuron cell is at rest it has a high concentration of potassium ions on the inside and a low concentration on the outside; and a low concentration of sodium ions on the inside and a high concentration on the outside. Ions passively flow through selective ion channels for potassium and sodium, and are actively transported by sodium-potassium pumps through ion channels until a proper balance is reached. The balance is called the resting potential. In neurons, the proper balance results in a resting electrical charge across the cell membrane of negative 70 millivolts (-70mV).

Resting potential is really misleading, for the neuron cell, like all other cells, must constantly expend energy through work in order to maintain the resting potential of -70mV, by actively transporting ions through sodium-potassium pumps imbedded across the plasma membrane. Thereby 3 ions of sodium are pumped out of the cell for every 2 ions of potassium pumped in.

When a neuron reaches its resting potential state of -70mV, it is ready to 'fire'. Scientists refer to this as reaching its action potential. It is now ready to send an electrical signal.

An action potential occurs when a neuron sends information down the axon, away from the cell body. The action potential is actually an explosion of electrical activity that is created by a depolarizing current. Here's what happens.

Depolarization occurs when some 'excitatory stimulus' causes the voltage-gated sodium channels in the cell's membrane to open wide and allow sodium ions to rush headlong through the channels down their electrochemical gradient. If the stimulus is strong enough to cause the membrane voltage to drop to -55mV, more and more voltage-gated sodium channels spring fully open and allow sodium ions to gush into the cell.

When depolarization reaches about -55mV the neuron 'fires' an action potential. This is an 'all or nothing event'. If the stimulus results in a drop from -70mV to -55mV the neuron 'fires'. If depolarization does not reach -55mV, nothing happens. So, the 'threshold' is always -55mV. And, the size of the action potential is always the same. Either the neuron reaches the threshold and a full action potential is fired, or it doesn't and no signal is sent. If the threshold is reached, the membrane voltage instantly raises to a peak of about +45mV. At that peak point, the voltage-sensitive sodium gates slam closed and the voltage-gated potassium channels open up. As potassium ions then begin to rush outside, the reverse occurs as the voltage within the cell falls back into negative values. This process is completed as the sodium-potassium pumps are actively employed to regain the resting potential of -70mV inside the cell. The neuron is now ready to fire again if stimulated. The time necessary for this wondrous process of neuron 'firing' to be accomplished is measured in milliseconds.

The depolarization process from -70mV to +45mV to -70mV again represents the generation of an action potential. The action potential or 'firing a neuron' thereby sends an electrical impulse charge (current) down the entire length of the axon transmission 'wire' until it reaches the end of the cell's axon terminal. The electrical impulse signal that is sent forth as an electrical wave down the axon 'wire' is called an 'action potential'. It has been generated when the polarized cell's 'resting potential' becomes depolarized as the result of being 'excited' by some stimuli, like touching, or stretching, or a chemical transmitter. More often than not the exciting agent is a chemical transmitter. And, that chemical transmitter usually originates when the action potential electrical impulse signal reaches the end of an axon terminal, which is adjacent to a dendrite on the next neuron.

When an electrical impulse signal reaches the end of an axon terminal it encounters a gap between the terminal end of the cell and the dendrites of an adjacent neuron. This gap is called a synapse. The electrical impulse signal has run out of 'wire' and cannot be electrically transmitted any further. However, the electrical transmission story does not end there. The electrical signal is then converted to a chemical signal, a most remarkable event."

* * *

Noah Tall paused for a few moments to catch his breath. After taking a few sips of water he continued.

* * *

"After traveling the entire length of the axon and reaching a synapse at the end of an axon terminal, the action potential (electrical signal) causes the release of chemical transmitters at the end of the axon terminal. These chemical neurotransmitters are released at an axon terminal when the action potential (electrical signal) opens voltage-gated calcium ion channels, thereby allowing calcium ions to enter the axon terminal. The calcium causes neurotransmitter molecules to fuse with the cell's membrane and then be released outside of the membrane to activate receptors in the dendrites of the neuron next door, adjacent to the synapse.

Chemical neurotransmitters released by other neurons at the end of their adjoining axon terminals may join forces to depolarize the neurons next door to the threshold level of -55mV. If the combined stimuli reach threshold for the next-door neurons, they induce a further action potential 'firing' in those neighboring neurons. That may, in turn, serve to propagate the electrical signal on down the line in the next neuron cell. And so on. And so on.

In this manner, a change in a neuron's state, from resting potential to action potential, usually occurs as the result of stimulation by a chemical neurotransmitter at the synapse of a nerve cell.

It must be noted that the wiring system in our nervous system is both quite complex and well insulated. Neurons have a myelinated cell membrane. The cell membrane is encased in a special protein coating called myelin. This myelin sheath both protects the cell and inhibits the flow of ions between the fluid within the cell and the fluid outside the cell. It also insulates the membrane, thus allowing the electrical signal to be transmitted nearly instantaneously and with virtually no loss of signal strength.

So, that is a nutshell version of how neurons work. I hope that is helpful."

Q: "Believe it or not, Professor, I'm almost at a loss for words. That is simply incredible.

But, I'm sure you know that I do have a most pointed follow-up question to ask?"

A: "I expect that I do. You would like to know how neurons learned how to do this amazing transmission of information without any direction through the process of evolution by the natural selection of random mutations of genetic DNA. Well, I'll be happy to tell you.

As we explain in our 2008 book, *Science, Evolution, and Creationism,* random sequences of genetic DNA mutations act as evolutionary experiments. Obviously when the random sequences produced neurons that could transmit information through electricity, that adaptive trait proved very useful to species survival. Again, scientific evidence for that is the fact that neurons occurred quite early in the evolution of animal life and have been highly conserved because they are of great benefit to the survival of animal species."

Q: "And, I had a feeling that would be your answer. That answer is again based on no confirmable scientific evidence. You observe that neurons operate in the exquisite manner that they do and then conclude that they **just had to** evolve through the process of **random** mutations of genetic DNA.

When scientists first began to observe living cells through microscopes they were absolutely convinced that they actually saw a tiny, tiny preformed person residing within a sperm sex cell of the father. They truly believed that a fully-formed baby simply grew larger and larger within the mother's womb until birth. They observed what **just had to** be true even though it belied all of the wonders of embryological development that we know to be factually true today. Those scientists saw what they wanted to believe was true based on their preconceptions.

Dr. Tall, isn't it possible that when you observe neural development and actually see **randomness** that your randomness preconception is clouding your objective judgment?"

A: "No. Of course not. The only scientific explanation that is possible for neural development is random mutation of genetic DNA."

Q: "Yet again, for the record, let me ask this question with regard to the evolution of neurons by the natural selection of genetic mutations. Which of the following explanations is the more scientific to present to youngsters in public school science class?
- The mutations were undirected and random.
- The mutations were undirected and random or they were directed and purposeful and no one knows which is true."

A: "Mr. Darrow, my answer remains the same and always will. The mutations were undirected and random. That is the only acceptable answer in science class."

* * *

Judge Raulston at this point called for a recess of the trial. After the recess Clarence Darrow continued. Goody Spyer had suggested during recess that it would be a good idea to next examine the evolution of two other wonders that he found from reading the lawyer's briefing papers to be every bit as compelling as the wonders of electric charge and information transmission by neurons.

* * *

Homeostasis Keeps Things In Balance

Q: "Professor, another wondrous feature of animal life is that of homeostasis, the ability to regulate and maintain a stable internal environment. Did homeostasis evolve early-on in the evolution of animals?"

A: "Actually, no. It started before animals evolved. Homeostasis is another example of a 'highly conserved' adaptive trait. Before there were any animals, plants developed homeostatic controls. In fact, within the last five years research scientists have gained insights into transition metal ion homeostasis in bacteria. So, science has found that homeostatic regulation even in single-celled organisms is critical and has been around for literally billions of years. As organisms further evolved the homeostatic systems became more and more complex."

Q: "Could you describe what types of things the homeostatic systems serve to keep in balance?"

A: "Of course.

Plants have evolved the ability to regulate such functions as seed germination and flower and fruit development. Various hormones have developed that allow the plant to actually monitor its external environment and to thereby sense the right time for functions to occur. Seed germination is a good example.

A plant seed will stay in a condition of dormancy until its hormones signal that the external conditions are right. Only when the plant determines that there is adequate light, water, temperature and soil nutrients will it germinate and sprout. If the proper conditions are not present a seed may lie dormant for a long, long time until its hormones signal that the conditions are okay. Mimosa seeds have been documented as germinating after a dormancy period of two centuries.

Animals have evolved extremely accurate and sensitive control systems. All animals regulate their blood glucose levels. Mammals regulate their blood glucose by releasing insulin and glucogon from the pancreas, as needed. And, the kidneys remove excess water and salt from the blood.

Our endocrine system secretes over fifty hormones that act as chemical messengers to allow for homeostatic regulation. The endocrine system glands serve to regulate our body temperature, our thirst, and our hunger as well as our blood sugar balance."

Q: "How does the endocrine system actually work to keep these things balanced?"

A: "For each variable being regulated, like blood sugar balance or temperature, all homeostatic control mechanisms have at least three independent components for the variable being regulated:
- A control center (like our brain) sets the range within which a variable must be maintained (like our body temperature that must be maintained within a very narrow range or else we will die).
- A stimulus relays changes in the environment.
- A receptor senses the stimulus and sends information to the control center.
- An effector, like muscles or organs, receives a signal from the control center to make the adjustments necessary to maintain a narrow-range homeostatic balance. After receiving the signal a change occurs to correct the deviation by using either positive or negative feedback mechanisms."

Q: "Professor Tall, that is amazing. The regulatory controls seem to be quite sophisticated and require a lot of coordinated activity to first set the desired parameters and then to monitor and adjust to maintain a homeostatic balance using both positive and negative feedback controls.

Did all of these regulatory mechanisms needed to maintain a balanced internal cellular environment evolve by natural selection of **random** genetic mutations without any intention or direction?"

A: "Of course they did. We have covered that point many, many times. That is the scientific explanation."

Q: "Is there any confirmable scientific evidence upon which the inference of natural selection of random mutations can be based?"

A: "Certainly. We have covered that many times. Natural selection of random genetic mutations is the only possible explanation if we are to remain scientific."

Q: "The answer that the evolution of sophisticated homeostatic regulatory systems in living creatures **just had to** occur randomly is reminiscent of the scientific belief that was maintained until after the start of the 20th century that the heavens **just had to** be filled with the mysterious substance called 'ether'.

'Ether' was the 'scientific explanation' provided by Aristotle in the 4th century BC.   Until Albert Einstein discovered the Theory of Special Relativity, the wave theory of physics required a medium for the propagation of waves. Our most brilliant scientists then believed that electromagnetic energy of the Sun **just had to** travel through the medium of Aristotle's 'ether'.   Einstein explained and proved that 'ether' was just another 'science fiction'.

Professor Tall, when you see randomness as the mechanism that **just has to** be true for producing exquisite, coordinated homeostatic regulatory systems isn't that a belief that is akin to the earlier scientific belief in the 'scientific truth' of heavenly 'ether'? Isn't randomness simply another 'science fiction'?"

A: "Of course not. Mr. Darrow, you certainly know how to be most insulting."

Q: "Professor, absolutely no insult is intended. I am simply trying to discover the truth.

Dr. Tall, which of the following explanations concerning the evolution of homeostatic mechanisms by natural selection of genetic mutations is the more scientific to present to youngsters in science class?
- The mutations were undirected and random.
- The mutations were either undirected and random or directed and purposeful and no one knows which is true."

A: "Yet again, Mr. Darrow, my answer is undirected and random. That is the only possible scientific answer."

The Immune System – Kill the Bad and Spare the Good

Q: "Professor Tall, it is my understanding that the immune system is another one of those 'highly conserved' adaptive traits that evolved quite early in the evolution of life. It that true?"

A: "Yes. Both plants and all animals have an innate immune system."

Q: "How does the innate immune system work?"

A: "The innate immune system is a non-specific defense against pathogen invaders that we share not only with other mammals but with plants and all other animals. While non-specific to invading pathogens, the innate immune system does depend on the ability of the system to distinguish between self and non-self molecules. It must kill the bad (non-self) but spare the good (self).

Immune cells are white blood cells known as leukocytes.

At the site of infection or inflammation from a wound, specialized chemical signaling agents, called cytokines, are called into action. Cytokines signal for help to certain leukocytes called phagocytes. Phagocytes are white blood cells that constantly patrol the body looking for pathogens. When a phagocyte then encounters a pathogen it kills it by eating it. The pathogen is engulfed within the body of the phagocyte cell and then killed by its cellular digestive enzymes.

Other leukocytes, known as mast cells, reside in mucous membranes and attack pathogen invaders associated with allergic reactions. When activated, these mast cells release histamines, which dilate blood vessels and signal phagocyte specialists to come to the site and kill the invaders.

Yet other leukocytes, known as natural killer cells, attack and kill host (self) cells that have been infected by invading pathogens. Natural killer cells accurately distinguish between healthy and infected host cells, then automatically kill the bad while sparing the healthy.

In general, the white blood cells of the innate immune system prevent the growth of many harmful bacteria within the body. However, many pathogens have developed the ability to evade the innate immune system. So, we have evolved another major line of defense against pathogen invasion that is present only in mammals, the acquired immune system."

Q: "Could you briefly describe how the acquired immune system operates?"

A: "Yes. And, I'll be very brief.
    In essence, we have specialized cells that circulate through both the blood and tissue fluids that build antibodies. The antibodies that are constructed by these specialized B cells and T cells are specialized proteins who kill invading pathogens by tightly binding and engulfing them. But, not all of the antibodies produced by the specialized B cells and T cells are used to kill the invaders. Some are retained as memory cells that are used to ward off potential future invasions from that specific pathogen. That is how we develop immunities to various diseases.
    If that unique pathogen is again encountered at some future time in our lives the memory cells will become active and secrete literally millions of copies of the antibody required to kill the specific pathogen by devouring its cells. Such represents the **acquired** immunity of this defensive system.
    The acquired immune system is able to distinguish between many different antigens. The receptors that allow for such differentiation are produced in very large numbers. The human body is capable of producing in excess of one trillion different antibody molecules. So an amazing degree of differentiation is required. And, the most important differentiation feature of the acquired immune system is the same as for the innate immune system: each must be able to accurately distinguish between self and non-self molecules in the body. Otherwise, the antibody response that kills the invading pathogen would kill us as well.

It is important to recognize that the innate immune system and the acquired immune system are part and parcel of a total package. They work together. The innate system would be swamped by rampant infectious agents without the acquired system. On the other side of the coin, the B cells and the T cells of the acquired system could not become functionally active in the first place without the help of the innate system. Together they protect us from a multitude of invading pathogens trying to kill us."

Q: "Dr. Tall, again, that is an amazing description. You have described a combined immune system that seems to be quite sophisticated and requires quite a bit of coordinated cellular activity. The ability of these tiny cells to be able to accurately differentiate which other cells belong to the organism itself and which are non-self invaders is absolutely amazing. And, the . . ."

A: "Mr. Darrow, even before you ask the question again that is inevitably coming, I will answer it. Yes. Every part and function of both the innate and the acquired immune systems evolved by the natural selection of random genetic mutations. That is the only explanation possible if we are going to remain scientific. And, yes, that is the only answer that should be taught in science class."

Q: "Professor, you have certainly provided a clear pattern in your answers so far today. Yet, I will continue to seek actual confirmable scientific evidence in support of your answers. So far I have found none.

You tell us that when you observe the wondrous ability of both the innate and acquired immune systems to protect us from organisms trying to kill us, that you also observe that these exquisite protection mechanisms were the product of randomness. You maintain that such is the only possible scientific answer.

In the not-too-distant past other brilliant scientists observed the spontaneous generation of life from non-living matter, whereby frogs grew out of mud, rats grew out of garbage, and flies were born from rotting meat. They observed these things and explained that they **just had to** be true. We now know that those explanations were entirely false. Those explanations were the result of the judgment of those scientists being clouded by their preconceptions.

Dr. Tall, you tell us that you firmly believe that the randomness explanation is the 'scientific truth' for the development of the immune systems. Couldn't your judgment be clouded by your preconception that randomness **just has to** be the answer?"

A: "No, of course not. The randomness explanation is simply the only possible scientific explanation. Every scientist knows that. I don't know why you simply cannot accept that scientific fact."

Evolution of Complex Body Parts

Q: "Professor, a final area I'd like to cover today is the evolution of complex body parts in animals. I'm talking about complex body organs like the heart and the eye and the brain, as well as complex biological systems like blood clotting.

The proponents of teaching Intelligent Design in biology class claim that such complex biological structures and systems are irreducibly complex and therefore could not have evolved in a step-by-step natural evolutionary process.

In your 2008 book *Science, Evolution, and Creationism* you refute that Intelligent Design claim with the following argument:

> 'However, the claims of intelligent design creationists are disproven by the findings of modern biology. Biologists have examined each of the molecular systems claimed to be the products of design and have shown how they **could have arisen** through natural processes.'

The explanation that then follows in that book first says that some protein components of organ structures are likely the precursors of other proteins. It then states that proteins which perform different functions can be similar in structure, and that this similarity indicates a common evolutionary origin.

The NAS textbook explanation then continues:

> 'Evolutionary biologists also have demonstrated how complex biochemical mechanisms, such as the clotting of blood or the mammalian immune system, **could have evolved** from simpler precursor systems.'

With that long prelude, Professor, let me ask this question. Is the explanation that complex bodily organs **could have arisen** and that complex biological systems **could have evolved** from simpler precursors supported by confirmable data whose results can be observed, tested, replicated and verified?"

A: "Of course it is or we would not have included that explanation in our book. The scientific evidence relies on the fossil record, the homology of anatomical structure, and genetic analysis of the DNA in protein families."

Q: "Professor Tall, is that same process responsible for the development of the human brain? You seriously believe that even the human brain evolved through a step-by-step process of nature selecting mistakes and accidents in genetic DNA, and that such a belief should be taught in science class?"

A: "Absolutely. The evolution of the human brain was the result of evolution by natural selection of random mutations. I know you have a hard time understanding that, Mr. Darrow, but that is the scientific truth."

Q: "I understand your answer. I simply find your answer to be entirely unsupported by any confirmable scientific evidence whatsoever.

Let me ask you this. Have scientists ever been able to construct a computer that is as efficient or as effective as the human brain?"

A: "No, not yet, but we have come a long way. Most people forget how new the computer age really is. The genius mathematician, Alan Turing, first developed the concept of digital computers in the 1940's. He described a computer that read a code of ones and zeros and he believed that science could create intelligent machines following the blueprint of the human brain."

Q: "But, scientists have not been able to do that, have they? In 1946 Alan Turing made this prediction:

> 'In 30 years it would be as easy to ask a computer a question as to ask a person.'

Brilliant scientists have been trying to develop such a computer for more than 60 years now and have not been able to do that. Yet still, your testimony is that mistake upon mistake in genetic DNA has produced the materials and refinements to evolve a human brain by accident, without any purpose or intention or direction?"

A: "The human brain evolved through the same process used to evolve all other adaptive traits of living organisms. The process is simply nature selecting useful random mutations of DNA, with no need for purpose or intention and with no direction in mind. I wish you could start to understand that.

Indeed, we have now developed a man-made computer that can process information in a quantity and speed approaching that employed by the human brain.

IBM's BlueGene is a mega-computer that can process 10 quadrillion bits of information per second. And, we have calculated that speed to be nearly identical to that of the human brain."

Q: "But, there are incredible differences in the nature of operations processed by that computer compared to the human brain, are there not? That computer has a highly-structured executive control center while human brains use a parallel architecture. Our minds continue to function and sort things out even when the information is incomplete, while computers are always one line of code away from freezing up. That's a huge difference.

Yet, the most incredible difference is one that we have seen before in observing how dumb cells figure out elegant solutions to intractable problems involving heat and electricity.

BlueGene is a complex of some 120,000 processors linked together. It uses 1,500,000 watts of power per second. That is roughly equivalent to the amount of energy required to supply the energy needs for 1,200 households in America. The human brain processes the same amount of information using only 10 watts of power. In other words, a man-made computer needs 150,000 times as much energy to perform the same number of operations as a human brain. Of course operating the brain with that much energy would kill us.

Professor, we are back once again to this conundrum. How did dumb cells figure out how to supply the energy required to operate the world's most highly sophisticated computer without burning up?"

A: "Mr. Darrow, we have already discussed how neurons operate. The brain uses an electrochemical process that simply doesn't require as much energy. It simply developed that process of parallel electrochemical neuron linkages by the natural selection of random genetic mutations that proved useful."

Q: "Dr. Tall, I understand your answer, but what is the specific confirmable data to support the inference that the brain evolved in that fashion?"

A: "I have already told you that. You seem to be incapable of understanding. The confirmable scientific data supports the clear finding that the brain and all of the body's organs and biological systems evolved by the natural selection of random genetic mutations. Period."

Q: "So the NAS scientific explanation is that natural selection of **random** mutations of genetic DNA is the mechanism for the evolutionary development over time of complex organs like:
- hearts
- lungs
- gills
- blood
- roots
- stems
- shelled egg
- mammal placenta
- kidneys
- liver
- eyes
- ears
- skin
- pancreas
- spleen
- stomach
- intestines, to name but a few;

and also complex internal systems like:
- blood circulatory system in mammals
- xylem and phloem circulatory systems in plants
- hormonal regulatory systems in plants and animals
- blood clotting system
- immune systems in plants and animals;

and finally the organs that make us uniquely human:
- a larynx and epiglottis that give us the power of speech
- a human brain that gives us the power of reflective thought?"

A: "That may seem incredible to you, Mr. Darrow, but that is the rock-solid scientific explanation."

The Power of Scientific Paradigms

Q: "Dr. Tall, it is quite obvious that scientists have discovered a great deal about how each and every one of these amazing physical organs and bodily processes works. And, we all owe a great debt to brilliant scientists for these discoveries. But, discovering how something works is not the same thing as understanding how it came to work the way it does in the first place. Let me ask you this.

If a scientist discovers how DNA is configured and coded to provide the information necessary to construct and operate each organ and system within each living organism he should be awarded the highest scientific award for that achievement – the Nobel Prize. Isn't that true?"

A: "It certainly is true and that is why Crick and Watson received that award."

Q: "Each such discovery that is recognized by a Nobel Prize is the result of both diligent hard work and scientific genius by brilliant scientists. Isn't that true?"

A: "Since I have also received a Nobel Prize, it would be egotistic to say that, but you certainly may draw that conclusion if you wish."

Q: "Yet, you and the vast majority of the elite scientists of the NAS firmly believe that each such configuration and coding and construction and operation that is discovered by brilliant scientists developed without either hard work or scientific genius. In essence, you believe that each configuration and coding and construction and operation of each living organism was the result of **random** changes in the nucleotide sequencing of genetic DNA.

No intelligence whatsoever was required to actually do these amazing things, but genius intelligence was required to simply discover them.

Is that your testimony?"

A: "Mr. Darrow, no intelligence was required for the evolution of living things. That may still seem amazing to you. However, you certainly are no scientist."

Q: "Professor, how can you be so certain that the changes in genetic DNA were random and undirected?"

A: "That is, quite simply, the only explanation possible if we are going to remain scientific."

Q: "Dr. Tall, scientists haven't always been so certain that changes in living things were random and undirected, have they?"

A: "No, of course not. For ages what was called science was filled with superstition and nonsense."

Q: "In fact the greatest scientific genius of ancient Greece, Aristotle, believed that all things resulted from four causes, and the so-called final cause was the **purpose** that the thing served. Aristotle's science was not based on randomness but, rather, on purposefulness. And his views were held by elite and mainstream scientists for a long, long time. You may call that superstition and nonsense now, but it was certainly believed to be true at the time, wasn't it?"

A: "Yes. But that makes it nothing else but superstition and nonsense."

Q: "A lot of brilliant scientists firmly believed in those things that you now call superstition and nonsense. Indeed, two fairly well-known scientists never gave up their belief in the ultimate reality of purposefulness: Isaac Newton and Albert Einstein.

Let me review, if I may, some of the history of science to illustrate how the power of group-think and paradigms have long-served to establish 'scientific truth'.

Aristotle developed an Earth-centered model of the universe. The planets in the heavens were believed to be perfectly circular bodies that traveled around the Earth in perfectly circular orbits. Aristotle explained that the empty space in the heavens above was filled with a mysterious 'ether'. Aristotle's Earth-centered universe was held to be the scientific 'truth' for more than 2,000 years. Isn't that true?"

A: "Yes, that is true. But the path of science was used to overcome that superstition and nonsense as scientific discoveries were made. On his deathbed in 1543 Nicholaus Copernicus published his book that explained mathematically that Aristotle was wrong. He explained that not the Earth, but the Sun, was the center of our solar system. Then after the telescope was invented in the 17th century Galileo Galilei made actual observations that comported with the Sun-centered solar system."

Q: "But history tells us that scientists went to great lengths to maintain the scientific 'truth' of an Earth-centered universe even as they discovered things that did not comport with such an explanation.

When scientists observed a retrograde motion of planets that did not follow the required perfect-circle pattern required by Aristotle, they simply made the new facts fit within the paradigm of an Earth-centered universe. They explained that such retrograde motion was in fact the result of smaller perfect circles moving within larger perfect circle orbits. They made the discoveries fit the paradigm. In fact, they did not give up belief in an Earth-centered universe until the evidence of a Sun-centered solar system became overwhelming.

Galileo was the first to use the newly-invented telescope to observe the heavens. He discovered sunspots and valleys of the Moon that were at odds with the Aristotelian notion of perfectly-formed heavenly bodies. He observed that the planet Jupiter had moons that circled it and that there were large variations in the brightness of the planet Venus, all compelling evidence that the Sun was the center of the solar system as Copernicus had first explained.

Yet, the elite and mainstream scientists of the time continued to hold fast to the 'truth' of an Earth-centered universe until well after the death of Galileo in the middle of the 17th century? Isn't that a fact?"

A: "Yes. But Copernicus had begun a scientific revolution. As the scientific revolution proceeded, old superstition and nonsense were laid to rest forever."

Q: "Yet certain features of Aristotle's universe had great staying power didn't they?

Until the end of the 19th century wave theory required that energy had to move through some sort of 'medium'. Ocean waves traveled through the medium of water and sound waves traveled through the medium of air. Since there was no water or air beyond the Earth's atmosphere, the solar energy of light from the Sun **just had to** move through some medium, for that is what wave theory required. Brilliant scientists held firmly to the belief that Aristotle's 'ether' was the necessary medium that filled the void of outer space until Albert Einstein revealed in his 20th century theory of Special Relativity that light did not have to travel through any medium at all.

My point is that history shows us that brilliant scientists have believed over and over again in the absolute **certainty** of things physical that later turned out not to be true. Isn't that a fact?"

A: "Yes. But the real brilliance of science is that we are always open to change based on new evidence that is discovered."

Randomness is the Modern Scientific Paradigm

Q: "Professor Tall, the nobility of science is indeed based on always being open to change in the light of new evidence. But, the problem with your overarching explanation of **randomness as the 'truth'** of evolutionary change is this. Randomness can never be proved **not** to be true. Randomness is an explanation that is absolutely incapable of being falsified. Isn't that true?"

A: "No, that is not true. If someone can provide confirmable scientific evidence that clearly demonstrates that something developed in fact not randomly, that it indeed developed through the mechanism of direction and purpose, then that evidence would falsify randomness."

Q: "But it is not possible to do that. Yet, on the other hand, no one can show that any feature of biology did not develop through the mechanism of direction and purpose, can they?"

A: "Indeed, that cannot be proven to be false scientifically. Anyone who wants to believe in creationism or Adam and Eve or Noah and his Ark are perfectly entitled to do so. Science can never 'prove' them to be wrong. However, science deals only in **natural** explanations. Supernatural things are beyond the purview of science. Direction and purpose are not natural explanations. The only possible natural explanation is randomness."

Q: "Your 'scientific truth' of the mechanism of randomness is based on the proposition that given enough time mistakes can do anything. That is not scientific. That is bull-headed. It ignores the fact that for changes in living beings we are not talking about infinity. We are talking about the 3 and ½ billion years that life has existed on Earth. That may be a really long time. But, the very laws of the science of probabilities tell us that it certainly is not enough time to accomplish all the wonders of living things if they, indeed, evolved through the process of natural selection of **random** genetic mutations.

My car is 10 years old. It is starting to burn oil and it uses a lot of gas. I need a new car with a more efficient engine that uses energy more efficiently. Let's proceed to produce such a new car by mistake.

A mistake is an accident. So, why don't I just have an accident that will re-arrange the old parts in my old car into a better machine? So, we run it off the road into a tree. Dr. Tall, no one would ever believe that if we did that a lot of times – a billion, billion, billion times – that I would ever get a new car. No, we would end up with a lot of junk and no trees on the planet. Yet, that is the very process that you insist must be used for all adaptive change to be made in living organisms.

Even our most brilliant scientists, filled with intention and purpose, cannot bring about the changes that you insist that living cells simply developed by accident. Remember the process of cellular respiration that produces 38 ATP energy molecules from each glucose molecule that is converted? That cellular process achieves a 40% efficiency rate while our Detroit automotive engineers can produce an engine that is at best 25% energy-efficient.

I guess cellular mistakes simply outsmart human intelligence every time. Is that correct?"

A: "Mr. Darrow, not only has this whole line of questioning been unproductive, it has been very tiresome. I am here as an expert witness to provide relevant testimony in matters of science. I will not dignify your fantasy with an answer to that ridiculous question."

Q: "I apologize, Professor Tall. I guess I do get carried away at times when I can't seem to get you to see the incredulity of your position regarding the randomness explanation.

Can you give me any example of what the possible confirmable evidence would be to scientifically support the explanation that **something** occurred on purpose instead of randomly?"

A: "No. I cannot. And, don't you see, that is exactly why the only possible scientific explanation is randomness."

Q: "History shows us that scientists teach the 'truth' of things in science class that are indeed not true, until new evidence is discovered that contradicts such 'truth'. Since it is not possible to discover confirmable scientific evidence that the randomness explanation is not true, is it not indeed misleading to students in science class to tell them that the randomness explanation is **in fact** true?"

A: "No. That is not misleading at all. The only scientific explanation possible is randomness."

Q: " Let me ask you one final question for today, Dr. Tall. Which of the following explanations regarding the mutations of genetic DNA that produced all of the complex organs and complex biological systems that we see in the modern organisms existing today is the more scientific to present to youngsters in public school science class?
- They were undirected and random.
- They were either undirected and random or they were directed and purposeful, and no one knows which is true."

A: "The answer remains that they were undirected and random. That is the only explanation that has any place in science class."

* * *

Clarence Darrow turned and addressed the Judge.

"Your Honor, that concludes my questioning for today. Tomorrow I would like to explore the creation of life by random mutations."

Judge Raulston then adjourned the Court until the morrow.

"Nothing is so firmly believed as that which we least know."

Michel de Montaigne
French Essayist (1533 – 1592)

CHAPTER 8

The Trial – The Fifth Day
Creation of Life by Random Mutations

The atmosphere in the courtroom became animated as the fifth day of the trial began. There was a sense that the tension between Clarence Darrow and Noah Tall was leading up to an explosive point. The gallery attendees were beginning to place wagers among themselves as to how much longer a physical confrontation could be avoided.

Clarence Darrow assumed his position in front of the Plaintiff's table, placed his cane atop the table and leaned back. Noah Tall resumed his seat on the witness stand. Judge Raulston made quick work of attending to the procedural details and called on Clarence Darrow to resume his questioning.

* * *

Q: "Good morning again Professor Tall. Again, I hope you had a pleasant evening.

In the last few days we have covered a lot of ground about how life on Earth evolved. Today I'd like to explore how life on Earth began.

You have told us that the first known life on Earth was a type of bacteria that contained DNA and many proteins that had to be constructed in the manner specified by the bacteria's DNA.

Could you tell us how many proteins must be constructed in the manner specified in the DNA code in order for a single-celled bacterium to function as a living organism?"

A: "The smallest genome for a living creature that has been discovered by science to date is that of tiny single-celled bacteria that live within the cells of a host insect. The genome of those bacteria contains about 160,000 base pairs of DNA coding for 182 discrete proteins. These bacteria cannot live independently. There are specific genes that the bacteria's genome lacks that are necessary for independent life. The insect host compensates for those genes necessary for independent life.

Science has calculated that the simplest independent single-celled living organism must have at least 200 genes coding for at least that many proteins in order to survive. So, that single-celled bacterium must have constructed at least 200 discrete functioning proteins."

Q: "Each one of those proteins would have been constructed through evolution by natural selection?"

A: "Yes."

Q: "And, you told us that in order to be alive that single-celled living organism had to be capable of doing a minimum of three things. It had to:
  • store and process information;
  • acquire and use energy; and
  • reproduce its cells and itself.

You explained that all living creatures contain DNA that provides instructions for both how the organism obtains and uses energy and for how it reproduces. Is that basically correct?"

A: "Yes."

The Complexity of All Living Things

Q: "A scientist friend of mine made the off-hand remark that those first living organisms, which you have explained were made by random mutations, were far more complex than the most complicated machine that mankind has ever been able to produce. Do you agree with that remark?"

A: "We, of course, don't know the precise structure of the first living organism. But, it is a reasonable conjecture that they were pretty similar to species of bacteria that exist on Earth today.   And, it is a factual statement that we have never developed a machine from non-living materials that is as complex as a living single-celled bacterium.   I would add, however, that I am quite confident that science will someday discover how to develop life from non-life."

Q: "Just so I am clear.   Our most brilliant scientists, filled with purpose and intention have never been able to create life from non-living matter. Yet, you are quite certain that the first bacterial life originated as the product of an unimaginable number of mistakes that occurred in non-living matter, each mistake occurring without any purpose or intention.   Professor Tall, you are one of the most brilliant scientists in America.   Let me ask you most bluntly.   Does that really make sense to you?"

A: "The explanation that life originated from non-life by the process of natural selection of random mutations is the only explanation that comports with science.   That explanation makes perfect sense to me."

How Bacteria Acquired Energy

Q: "How did those first living creatures obtain the energy they needed for living?"

A: "They ate gas and rocks."

Q: "Could you elaborate a little more?"

A: "For roughly the first 500 million years of life on Earth, all living organisms were single-celled bacteria that contained no nucleus. They had the ability to obtain all the energy they needed for living directly from the elements of the Earth. These autotrophs possessed the ability to elaborate all of the organic compounds they required for their growth and reproduction directly from carbon dioxide in the mineral elements, which they had the ability to fix."

Q: "How did those earliest living organisms know how to obtain the energy they needed to live?"

A: "The information is contained in their DNA."

How Living Organisms Reproduce

Q: "How did those single-celled creatures reproduce?"

A: "They split in half."

Q: "I know that it must be tiresome to continue to provide more detailed explanations for things that are so simple to you. But, it would help greatly for our understanding if you could describe that reproduction process. Would you do that for us?"

A: "Yes. Of course. All cells reproduce by splitting in half. Since the first living organisms contained but a single cell, the reproductive process was performed by the single cell simply splitting in half. The same principle of cell reproduction applies for all cells. For simplicity of understanding I will use the example of one of our nucleated cells. Let me start with the first step, DNA replication.

   Just prior to cell division, whereby one living cell splits apart to become two living cells (a process called mitosis) the DNA double-helix, wound-up in the cell nucleus, is copied exactly.

   DNA is copied in such a manner as to exactly replicate the order of nucleotides along the polymer DNA strands through the complementary base pairing of the nucleotides with each other across the two strands of the double-helix of DNA. That process is, indeed, quite amazing.

   First, a protein enzyme called helicase attaches to a specific point on the double-helix, breaks the hydrogen bond between the two strands of DNA, and proceeds to unwind the double-helix, starting as a small bubble and proceeding outwards in both directions. Another protein enzyme holds the double-helix open. Yet another protein enzyme, called DNA polymerase, starts at the fork and simply continues to make a new and complementary strand of DNA from the template of the old DNA strand. The DNA polymerase that has attached to each strand of the DNA proceeds to copy the mirror image of the DNA strand simply by lining up and bonding the nucleotides in a sequence that creates the complementary nucleotide base on the new strand. As DNA polymerase completes a segment and moves forward, another protein enzyme causes the two strands to rewind together into the double-helix. Two identical DNA double-helix molecules have emerged from one. Each now contains one old strand of DNA and one new strand of DNA that forms a double-helix for each of the two identical DNA molecules.

Remember, all DNA is structured by a simple relationship of the matching of A with T nitrogenous bases and C with G nitrogenous bases in alternating pairs of nucleotides. In this manner, the nucleotides line-up in the sequence necessary for accurate replication of the entire DNA structure. The DNA replication process is not completed until DNA polymerase and yet other protein enzymes conduct tests for accuracy and correct mistakes. The result is an error rate in DNA replication that is less than one mismatch in over 1 billion nucleotides. After DNA replication has taken place, the nucleus of the cell now contains two identical complete sets or copies of the chromosomes containing the DNA for the organism.

In the life of a cell, once DNA replication has taken place, the cell is ready to divide in two. The division process begins with the membrane surrounding the nucleus starting to break down. As the nucleus membrane disintegrates, the chromosomes containing the DNA condense and then the two copies of the chromosomes (called chromatids) physically migrate to the center-line of the cell. A special protein attaches one of the chromatids to the top of the cell and the other chromatid to the bottom of the cell by a kind-of guy-wire arrangement. The two sister chromatids are thereby pulled apart with one on each half of the cell. The cell then is cinched-in along the centerline until it physically splits in two. Two daughter cells have now been created where only

one existed before. Each of the daughter cells now has the
exact sets of chromosomes containing the exact DNA that the
original cell contained.

Throughout the process of embryo genesis and then
throughout the life of the new offspring child, when new cells
are needed to perform specific functions this process of cell
division, called mitosis, will be employed to make exact copies
of the parent cell. Each new cell will thereby contain and
be able to extract the specific information necessary for the
proper functioning of that cell from the DNA library residing
in the cell's nucleus."

* * *

Noah Tall stopped his testimony for a few moments, consulted
some notes before him, and continued.

* * *

"For all multi-celled organisms, for all plants and all animals,
including us, the creation of a new living organism occurs through
sexual reproduction.

Sexual reproduction occurs when one of the special sex cells
(sex cells are called gametes) of the mother (the egg) is fertilized
by one of the special sex cells of the father (the sperm). The
methods of fertilization vary widely between plant and animal
species (e.g., sexual intercourse for mammals and pollination for
seed plants), but the result is always the same - a fertilized egg,
called a **zygote**.

Each of the cells of a multi-cellular organism, except the
sex cells, has tightly-wound within its nucleus the DNA of that
organism contained within a number of **pairs of chromosomes**.
In contrast, the sex cells, called **gametes**, contain only one-half
of each of the chromosome pairs. The process whereby the new
zygote cell receives a complete paired set of chromosomes (half
from each parent) is called **meiosis**.

Through meiosis the sex cells join together to form a single new and unique living cell - a **zygote**. Only the sex cells of the father, gametes called sperms, and the sex cells of the mother, gametes called eggs, contain but a single copy of each of the DNA chromosomes. The sex cells are produced by an unwinding of the DNA into single strands. Only one strand of each of the chromosome pairs is supplied by the father. The other strand of each of the chromosome pairs is supplied by the mother. Through sexual reproduction the single strands are mated to produce a zygote. The result is a **recombination of chromosomes**, whereby the mother's gamete sex cell (egg) is fertilized by the father's gamete sex cell (sperm) to become a new living organism's first cell. The first single cell of new life, the zygote, thereby contains DNA chromosomes that are double-stranded, with one strand each coming from the father and one strand each coming from the mother. The DNA of the new living zygote single cell is thereby unique. **A new unique life has been created to begin as a single living cell.**

Evolution, of course, is a step-by-step development of living organisms. The processes of mitosis and meiosis were essentially combined in the first living single-celled organisms without a nucleus. As I said at the outset. The first living organisms reproduced simply by splitting in half."

Q: "As usual, Professor, your explanation is detailed and fascinating. Thank you. But, back to my recurring point. How did the first living organisms know how to do all that?"

A: "Once again, and always, Mr. Darrow, the required information is in their DNA."

Where did the Information in DNA come from?

Q: "So, the information necessary to acquire the energy needed for life and the information necessary to reproduce itself was contained in the DNA of the first single-celled living creatures?"

A: "Again, yes. I've covered that."

Q: "How did that information get into its DNA? Where did that information come from?"

A: "The information in the DNA was acquired by a very natural process. The process was the evolution by natural selection of favorable mutations of the genetic materials of prebiotic structures, called proteinoids and purines and pyrimidines. Those materials were contained in protocells that were ancestral to the first living organisms that we know of."

Q: "When you say favorable mutations, what do you mean?"

A: "As genetic materials are reproduced copying errors sometimes occur. These copying errors, or mistakes, are called mutations. Mutations can also occur by an alteration of genetic materials in response to some radioactivity, cosmic rays, or some poison in the environment. When these mistakes or accidents occur they usually result in a neutral effect or a degradation. But, sometimes the mutation provides the organism with an advantageous trait that allows the organism to better meet environmental challenges. That's what I mean by favorable mutations."

Q: "And you are quite sure that these favorable mutations were random?"

A: "And, once again, yes. I am sure."

Q: "But, the prebiotic structures that you referred to, the proteinoids and purines and pyrimidines contained no genes, no genetic materials, did they?"

A: "Technically, no. But, the evolutionary process was the same. Random mutations occurred in the precursors of proteins and DNA and RNA that provided 'experiments' for evolution."

Q: "So, proteinoids were evolutionary ancestors to proteins and purines and pyrimidines were evolutionary ancestors to the nucleic acids DNA and RNA?"

A: "Essentially, yes."

Q: "You have told us that life requires three discrete things. In order to be alive:
- A living cell must be able to obtain, store, and process all of the information necessary to first bring the cell to life and to then provide for its every function.
- A living cell must be able to extract energy from its environment and then convert that energy into a useful form that is required to power the processes of cellular metabolism that are essential for all life.
- A living cell must contain the information and possess the ability to replicate. It must be able to clone itself, and the living organism must be able to produce offspring.

If proteins and DNA and RNA are the structures that are required for an organism to be alive, how could they have non-living evolutionary ancestors like proteinoids and purines and pyrimidines residing in some sort of protocells?"

A: "Perhaps it would be useful for me to simply read a pertinent section from our 2008 book entitled *Science, Evolution and Creationism*. The section explains quite succinctly how things began to come alive."

Q: "By all means. Please do so."

A: "This is the section.

'For life to begin, three conditions had to be met. First, groups of molecules that could reproduce themselves had to come together. Second, copies of these molecular assemblages had to exhibit variation, so that some were better able to take advantage of resources and withstand challenges in the environment. Third, the variations had to be heritable, so that some variants would increase in number under favorable environmental conditions. No one . . . . '"

Q: "Professor, let me interrupt for a moment before you proceed so that we may clearly understand. Let me ask a question or two about each of the three conditions you just read from the NAS book.

Has any scientist ever observed any non-living molecule reproducing itself?"

A: "Not yet. No."

Q: "Why would reproduced copies of these non-living molecules exhibit variation?"

A: "Because reproductive variation is an essential component of the process of natural selection."

Q: "Is there any scientific evidence in support of reproductive variations that are heritable in non-living molecules?"

A: "No. Not yet."

Q: "Why then do you call that explanation a scientific explanation?"

A: "Because that is the only explanation that comports with naturalistic evolution."

Q: "So, that is a scientific explanation because it is the only explanation that comports with random, undirected processes. And, you believe that such an explanation should be taught in science class even when there is no confirmable data in support of it?"

A: "Yes. It is a valid inference from the theory of evolution by natural selection."

Q: "Well, that clearly explains everything, doesn't it? Would you please continue with your reading from the NAS book?"

A: "Certainly.

> 'No one yet knows which combination of molecules first met these conditions, but researchers have shown how this process might have worked by studying a molecule known as **RNA.** Researchers recently discovered that some **RNA** molecules can greatly increase the rate of specific chemical reactions, including the replication of parts of other RNA molecules. If a molecule like RNA could reproduce itself (perhaps with the assistance of other molecules), it could form the basis for a very simple living organism. If such self-replicators were packaged within chemical vesicles or membranes, they might have formed 'protocells' - early versions of very simple cells. Changes in these molecules could lead to variants that, for example, replicated more efficiently in a particular environment. In this way, natural selection would begin to operate, creating opportunities for protocells that had advantageous molecular innovations to increase in complexity.'

Mr. Darrow, I think that should make things pretty clear, even for you."

Q: "One of my Mother's favorite sayings was: If ifs and buts were candy and nuts we would all sit down and have a party. The explanation that you just read certainly has a lot of ifs and buts and might-haves and coulds.

Do all of those ifs and might-haves and coulds actually provide a scientific explanation based on confirmable empirical data?"

A: "Mr. Darrow, you may choose to be insulting. But, our explanation is based on extensive scientific research performed by many dedicated scientists. It is rock-solid science. Just because we don't have all the answers yet doesn't mean that the explanation is not scientific."

Q: "Professor Tall, let me read again a very clear statement from the *'Only Science in Science Class Act'*. And I would note, by the way, that the exact language is contained in the NAS book entitled *Science and Creationism*. The unequivocal statement is this:

> 'Explanations that cannot be based on empirical evidence are not a part of science.'

Now I ask you again. What is the empirical evidence that supports the explanation you just read to us from the NAS book that the Dayton School District insists Mr. Scopes must use in his biology class? What is the empirical evidence?"

A: "Dedicated scientists have been working diligently for a long time to discover the specific evidence. Much progress has been made as I have already indicated. Scientists work on intractable problems and their work should be lauded, not disparaged just because we have yet to discover all the answers to very difficult questions. Science is a step-by-step process. Let me read another section from the NAS book, *Science, Evolution and Creationism*:

> 'Figuring out how life began is both an exciting and a challenging scientific problem . . . . Nevertheless, researchers have been developing hypotheses of how self-replicating organisms could form and begin to evolve, and they have tested the plausibility of these hypotheses in laboratories. While none of these hypotheses has yet achieved consensus, some progress has been made on these fundamental questions.

Since the 1950's hundreds of laboratory experiments have shown that Earth's simplest chemical compounds, including water and volcanic gases, could have reacted to form many of the molecular building blocks of life, including the molecules that make up proteins, DNA, and cell membranes. Meteorites from outer space also contain some of these chemical building blocks, and astronomers using radio telescopes have found many of these molecules in interstellar space.'

That, Mr. Darrow, is scientific evidence."

Q: "I am the first to admit that all of those hypotheses and experiments are laudable. The experiments themselves are scientific. They are designed to discover a natural process whereby life began by random, undirected, causes. But, the experiments have yielded absolutely no scientific evidence to support the hypothesis that life arose through purely natural causes, without direction or purpose. The fact is that no one knows how life began. Do you or any other scientist at NAS know how life began?"

A: "No, not with specificity. Not yet."

Q: "But, you are quite certain that life began simply by natural random and undirected causes, without any purpose involved. In your book, *Science and Creationism*, this unequivocal statement is included without any reservations:

'For those who are studying the origin of life, the question is no longer whether life could have originated by chemical processes involving nonbiological components. The question instead has become which of many pathways might have been followed to produce the first cells.'

Professor, do you really believe that proclaiming, without any evidence, that life simply arose from non-life by random, undirected natural causes is a scientific explanation?"

A: "I most certainly do. That is the only explanation that comports with the discipline of science. And, it is the only explanation that should be taught in science class. That is why the NAS included it in our book."

Q: "Do all scientists agree that life arose from non-life from purely natural causes?"

A: "All true scientists do."

Q: "Francis Crick is a renowned scientist who was awarded the Nobel Prize for discovering the structure and cryptic code in DNA. Is he a true scientist?"

A: "Of course."

Q: "In his book, *Life Itself: Its Origin and Nature*, Francis Crick was awed by the complexity required to begin first life. He then offered a solution in what he termed panspermia. He proposed that the first life on Earth could have occurred when bacterial spores were delivered to our planet, complete with the nucleic acids and proteins necessary for life.

Is that explanation that pre-made RNA and DNA and proteins were delivered ready-made to Earth a natural explanation?"

A: "Of course it is. As I explained earlier, meteorites from outer space have been found to contain some of the chemical building blocks for life, and astronomers using radio telescopes have found many of these molecules in interstellar space. It will not be surprising to scientists when we discover microbial life on the planet Mars. That would comport very well with Crick's theory."

Q: "But, doesn't that just beg the question of how those ready-made nucleic acids and proteins were developed on Mars or anywhere else in the universe?"

A: "No, it does not. Be it on Earth, or Mars, or anywhere else in the universe, life developed from non-life by the process of evolution by natural selection."

The 'Must Have Been' Explanation of Science

Q: "In the NAS book *Science and Creationism* you explain that bacteria are the earliest organisms that science has discovered in the fossil record. You then make this further unequivocal statement that even simpler life forms **must have** preceded those bacteria.

> 'These early organisms **must have been** simpler than the organisms living today. Furthermore, before the earliest organisms there **must have been** structures that one would not call 'alive' but that are now components of living things.'

Professor Tall, what is the scientific evidence in support of that explanation?"

A: "Mr. Darrow, apparently you failed to read further in that NAS book, or else you would know the answer. Let me read on:

> 'An important new research avenue has opened with the discovery that certain molecules made of RNA, called ribozyemes, can act as catalysts in modern cells. It previously had been thought that only proteins could serve as the catalysts required to carry out specific biochemical functions. Thus, in the early prebiotic world, RNA molecules could have been 'autocatalytic' – that is, they could have replicated themselves well before there were any protein catalysts (called enzymes). Laboratory experiments demonstrate that replicating autocatalytic RNA molecules undergo spontaneous changes and that the variants of RNA molecules with the greatest autocatalytic activity come to prevail in their environments. Some scientists favor

the hypothesis that there was an early 'RNA world', and they are testing models that lead from RNA to the synthesis of simple DNA and protein molecules. These assemblages of molecules eventually could have become packaged within membranes, thus making up 'protocells' – early versions of very simple cells.'

Mr. Darrow, that is the scientific evidence in support of that explanation."

Q: "Professor Tall, **could have** doesn't provide any empirical evidence necessary to qualify as a scientific explanation. And, 'protocells' have no place whatsoever in science. Indeed, there is no such thing as a 'protocell'.

Again, the referenced hypotheses and experiments are laudable. They are scientific. They are designed to discover a natural process whereby life began without purpose purely by random and undirected causes. But, the experiments have yielded absolutely no scientific evidence to support the hypothesis that life arose through purely natural causes, without direction or purpose. Isn't that correct?"

A: "The evidence is the natural process of evolution itself. In our book *Science and Creationism* we clearly explain how this explanation is scientific:

'The tremendous success of science in explaining natural phenomena and fostering technological innovation arises from its focus on explanations that can be inferred from confirmable data.'

Mr. Darrow, the confirmable data is that nature selects random mutations to evolve living things. There is an abundance of scientific evidence for random mutations. The inference that life began by the same natural process of the selection of random mutations that aid survivability is based on that confirmable data. That, Sir, is how science works."

Q: "And that answer, Sir, is what we lawyers call 'gobbledy gook'. That answer is made out of nothing but hole cloth. There is no scientific evidence whatsoever to support the explanation that life itself is the result of random mutations. But, you feel that it is most appropriate to teach youngsters in science class the two things about how life began that are included in the NAS book *Science and Creationism*. I'll repeat the explanations included in your book:

- There **must have been** organisms simpler than the ones living today (simpler than bacteria).
- Before the earliest organisms that lived on Earth there **must have been** structures that one would not call 'alive' but that are now components of living things.

In essence life **just had to** evolve from non-living molecules. Is that correct?"

A: "Yes. That is correct."

Darwin's God Works Miracles

Q: "Do miracles have any place in science?"

A: "Of course not."

Q: "How do you define a miracle?"

A: "A miracle is an extraordinary event which either cannot be explained by natural causes or which defies the natural laws of physics."

Q: "So when the Bible tells us that when Moses parted the Red Sea or when Joshua made the Sun stand still, those were miracles."

A: "Yes. Those are good examples of miracles. And, religion is filled with such miracles that cannot be explained by natural causes or that defy the laws of physics."

Q: "The God of Christians and Jews supposedly created life from the dust of the ground. Their Purposeful God did so intentionally. No natural law can explain how that extraordinary event could have happened. So, creating life from the dust of the ground is correctly called a miracle. Isn't that right."

A: "Yes. That would be a miracle. But, I don't know what all this talk about miracles has to do with science."

Q: "Well, let me take a moment to explain as the foundation for my next question.

The scientific explanation for how life began on Earth is that **somehow** out of the dust of the ground (the naturally-occurring inert inorganic molecules) organic molecules were **somehow** formed. Then those organic molecules **somehow** combined to produce a tiny single-celled creature that was alive. That tiny creature had **somehow** developed a complete coded language. That complete coded language then **somehow** provided the information necessary to allow the tiny creature to come to life, possessed of the ability to do the three things that are necessary for all life. And, that creature, so tiny that thousands of them together could not be seen by the unaided human eye, possessed the ability to do all these things, all at the same time:

- It had to be able to obtain, store, and process all of the information necessary to first bring it to life and to then provide for its every function.
- It had to be able to extract energy from its environment and then convert that energy into a useful form that is required to power the processes of cellular metabolism that are essential for all life.
- It had to contain the information and possess the ability to replicate. It must have been able to clone itself, and the living organism must have been able to produce offspring.

These extraordinary events, all occurring together at the same time, cannot be explained by natural causes. So science invokes the mechanism of **randomness** to provide the necessary explanation.

The Purposeful God of Christians and Jews invokes the **mechanism of intention** to explain the creation of life from the dust of the ground. The Accidental God of scientific atheists invokes the **mechanism of randomness** to explain the creation of life from the dust of the ground. There is no scientific evidence in support of either the mechanism of intention or the mechanism of randomness.

Professor Tall, isn't that the 'scientific truth'?"

A: "Mr. Darrow, I have testified repeatedly that there is abundant scientific evidence in support of the randomness explanation. That is the only explanation that is possible if we are to remain scientific."

Q: "Dr. Tall, let's examine the issue a little further, by reference to the field of science and mathematics that actually deals with the phenomenon of randomness. The science of statistics deals with the laws of probability or the laws of chance that provide the basis for the study of randomness. Isn't that correct?"

A: "Yes."

Q: "Quite reputable scientists believe that life was generated spontaneously by random events. George Wald, a member of the National Academy of Sciences and recipient of the Nobel Prize, made this observation:

'One only has to concede the magnitude of the task to concede the possibility of the spontaneous generation of a living organism is impossible. Yet here we are – as a result, I believe, of spontaneous generation . . . . Given so much time the "impossible" becomes possible, the possible probable, and the probable virtually certain. One has only to wait: time itself performs the miracles.'

Yet, science has discovered compelling evidence that the time for universal and Earthly development is not infinite. Science has discovered that the universe began roughly 14 billion years ago and that this planet was formed about 4 and ½ billion years ago. The laws of probability deal with finite things. The age of the universe and of Earth is finite. And, the science of statistics tells us that in our finite universe everything is **not** possible. The impossible does **not** become possible.

Noted scientist Carl Sagan stated that the odds for constructing a single protein of an average length of 100 amino acids by chance is greater than the odds of selecting but one particular atom from the total number of atoms in the entire universe. Noted statistician Emil Borel concluded that when odds of an event occurring become that highly improbable then the highly improbable becomes, in fact, **impossible**.

In short, the mathematical laws of probability themselves provide evidence that producing life through the mechanism of randomness is **impossible** through known natural causes.

Indeed, Francis Crick, who won the Nobel Prize for his discovery of the nature of DNA structure, provides this observation:

'An honest man, armed with all the knowledge available to us now, could only state that in some sense, the origin of life appears at the moment to be almost a miracle, so many are the conditions which would have had to have been satisfied to get it going.'

Professor, can you provide any scientific evidence that would contradict Francis Crick's observation?"

A: "I highly respect Dr. Crick and his views. I would simply observe myself that a great deal has been learned about the nature and functioning of DNA since his amazing breakthroughs and discoveries.

Again, my observation and my testimony is that the randomness explanation is based on solid science and it is the only possible explanation if we are going to remain scientific."

Q: "Professor, to your knowledge has any statistician ever computed the odds that would show that it is even possible to create life from non-life by chance?"

A: "No. But, I would add that I really don't know that many statisticians."

Q: "To your knowledge has any statistician ever computed the odds that would show that it is even possible for any of the following biological phenomena to have evolved by the process of nature selecting mutations that occurred by random chance?
  • Development of DNA information systems.
  • Development of photosynthesis.
  • Development and maintenance of electrical charge in living cells.
  • Development of enzymes to catalyze chemical reactions.
  • Development of an immune system based on the organism's ability to recognize 'self from non-self'.
  • Development of homeostatic controls to maintain internal environments within discrete parameters."

A: "Same answer."

Q: "If the odds can't even be computed for the occurrence of these miraculous phenomena by the mechanism of randomness, then the fact is that the randomness explanation cannot be used to support the hypothesis of natural causes. And, by your own explanation that would define these phenomena as miracles.

  The worshipers of Darwin's God believe that the mechanism of randomness first produced the miracle of life and then produced these other miracles by evolution through natural selection.

  Dr. Tall, isn't that the real 'scientific truth'?"

A: "Mr. Darrow, as I have repeatedly testified. I believe in science, not miracles. Would you like me to stop teaching science and start teaching miracles in science class?"

Q: "Professor Tall, I don't want you to stop teaching science in science class. I want you to stop teaching your religion of scientific atheism in science class. Can't you understand that?" That question, of course, was rhetorical.

* * *

Goody Spyer's animated comb-over was noticed by Clarence Darrow. The Plaintiff's attorney asked Judge Raulston for a recess, which was granted.

Goody advised Clarence Darrow that his line of questions and the elicited answers were apparently being well-received by the Judge. If he could continue to stress the actual religious nature of Noah Tall's explanations, then he would have a much better chance of showing that the School District's Policy actually was a violation of the Establishment Clause of the Constitution. Stay away from technical stuff for a while and stress the big picture.

When the courtroom session reconvened Clarence Darrow heeded Goody's advice.

* * *

Teaching the Truth of Science

Q: "Dr. Tall, you have noted several times during the course of this trial that I am not a scientist, trained in the ways of science. That is certainly true. I am a lawyer, trained in the ways of the law.

In the law we derive logical conclusions based on the strength of the actual evidence presented at trial. That is why I have repeatedly pressed you to provide actual evidence in support of your explanations.

Explanations not supported by evidence have no place in the courtroom or in the science classroom. I may personally believe in a Purposeful God. But, as a legal professional I fully understand and agree that the doctrine of Intelligent Design that supports that belief is not based on confirmable scientific evidence. My belief is therefore barred from being taught in science class as a violation of the Establishment Clause, and rightly so. Why can't you understand that the doctrine of randomness is a belief that is not based on confirmable scientific evidence and that your belief should therefore also be barred from being taught in science class as a violation of the Establishment Clause?"

A: "Because your belief is not based on scientific discoveries. My beliefs are. That is why your belief should be barred from science class but mine should be taught there."

Q: "Professor, it is impossible to falsify the explanation of randomness.

Yet, you insist that the mutations of the genetic DNA molecule that result in evolutionary change **just have to be** random. The effect of that is to make the randomness explanation a 'scientific truth' that simply cannot be wrong.

History has provided us with many such 'scientific truths' that simply had to be true but proved to be absolutely false, like:

- An Earth-centered universe to explain the workings of the heavens.
- A preformed tiny person simply growing larger and larger inside the mother's womb to explain embryonic development. Scientists truly believed that they actually saw a tiny fully formed person residing in the sperm cell. They believed that the father 'planted his seed' and the fully formed person just grew larger in the mother's womb until a baby was born.
- Spontaneous generation of life from non-living matter, whereby frogs grew out of mud, rats grew out of garbage and flies were born from rotting meat.

These 'scientific truths' have now taken their rightful place in the rubbish bin of science due to the discovery of confirmable scientific evidence by brilliant scientists. Don't you think that with all of the recent discoveries of science that confirm the existence of informational command and control functions within all living organisms that your explanation that organic change **just has to be** random will soon join those other 'scientific truths' in the rubbish bin of science?"

A: "No. I most certainly disagree. Randomness is an absolutely essential explanation for science. Science deals only with the material, natural world. Any explanation that is, at core, not based on randomness is simply beyond the purview of science."

Q: "Dr. Tall, you are the cream of the crop in science. The National Academy of Sciences is the most prestigious scientific association in this country. You are the head of the NAS. You are the elite of the elite, are you not?"

A: "Mr. Darrow, it is impossible for me to answer that question and remain self-effacing."

Q: "Well. You are. And history shows us that the elite scientists of their times have always insisted that the budding young science students **just had to** accept the parameters of study imposed by the elite.
   When Francis Bacon, the father of the scientific method, entered Cambridge University in the 16th century he was admonished that the university professors forbade questioning the 'truth' of Aristotle's science by their Charter:

> 'All students and undergraduates should lay aside their various authors and only follow Aristotle and those who defend him.'

The modern version would explain that the National Academy of Sciences has replaced Aristotle, admonishing:

> 'All students should only follow the randomness explanation of the NAS and those who defend that explanation.'

Isn't your insistence that the explanation of randomness **just has to be** true a simple update of the 16th century Cambridge admonition?"

A: "No, Mr. Darrow.   It is not.   In fact it is the opposite. Randomness is simply a parameter of science that is in fact set by the naturalism requirement of the scientific method itself."

Q: "If impressionable youngsters are taught by our most prestigious scientists that randomness is a 'scientific truth' for biological change then they will be disinclined to explore the phenomena of biological change any further. Isn't the nobility of science based on the door of exploration always remaining open or, at least, ajar?"

A: "Mr. Darrow, that question is personally insulting. I always encourage students to keep an open mind. That is what science is all about. I teach only the truth of science. Nothing more and nothing less."

Q: "Professor, I assure you that no personal insult is intended. I admire you greatly. But, when you teach the 'scientific truth' that all adaptive evolution is based on randomness, doesn't that teaching prove that there is no need for a Purposeful God?"

A: "I distance myself completely from any discussion of God or religion. Religion is not a consideration of a scientist."

Q: "Yet, your teaching of the 'scientific truth' of randomness has a compelling effect on the religious beliefs of your students, does it not?"

A: "Again, I wouldn't know. As I have said repeatedly, religion is not a consideration of a scientist."

Q: "Professor Tall, when impressionable youngsters are taught in science class that randomness is a 'scientific truth', they are taught that all living things, including themselves, developed by accident, without purpose and without meaning. Can't you see that by teaching them that doctrine that you have closed the door on the idea that there is any real purpose and meaning in life other than what we simply make up?"

A: "Mr. Darrow, that is not my concern. I leave the teaching of philosophy to others. My concern is simply teaching the truth of science."

Q: "Professor Tall, which of the following explanations concerning the origin of life on Earth is the more scientific to present to youngsters in science class?
   • Life originated from non-life without purpose, through random and undirected natural causes.
   • No one knows how life began."

A: "Life on Earth originated from non-life without purpose, through random and undirected natural causes. That is the only scientific explanation that should be taught in science class."

* * *

With that, Clarence Darrow turned and returned to the Plaintiff's table and sat down. Judge Raulston then adjourned the proceedings for that day.

CHAPTER 9

The Trial – The Sixth Day
Teaching Random Mutations in Science Class

What I observed in the courtroom at the start of the sixth day was palpable tension. The country way of putting it is 'You could cut it with a knife'.

Judge Raulston seemed to be most eager to proceed with the day. He obviously had been studying the matter with great intensity. He seemed anxious to witness the finale of this question and answer marathon on evolution by the natural selection of genetic mutations.

After the bailiff had called the session to order, the Judge called both Clarence Darrow and William Jennings Bryan before his bench. What I discovered later was that he had once again asked Mr. Bryan, counsel for the Defense, if he was sure that he wanted to continue to refrain from any cross-examination of the witness. Mr. Bryan again assured Judge Raulston that Professor Noah Tall was quite capable of presenting accurate testimony without any need of a lawyer's help. With that the trial continued with Clarence Darrow's questions and Noah Tall's answers.

* * *

Q: "Good morning Professor Tall. I hope you had a pleasant evening.

We have now reached the point where we get to examine the actual mechanisms of random genetic mutations that are taught as true in science class.

The central issue of this trial is, of course, the actual evidence that is presented in science class to support the explanation of randomness. You have testified repeatedly that only science should be taught in science class. And, you don't just say that random mutations are the change agents. You present actual evidence in support of that assertion. And, you can show us that evidence this morning, can you not?"

A: "I most certainly can. I would only ask that you try to stay to the point."

Q: "I will do the best that I can. I ask your understanding that some of this subject matter is not all that easy for laymen to digest quickly. If I seem to wander it is because I am trying to obtain as clear an understanding as possible. With that, I will proceed.

You have testified that in biology a mutation is a change in the information content of DNA. Could you elaborate on that?"

A: "Sure. The blueprint for all the information about each and every part of each living organism on this planet is contained in the DNA of the organism. DNA is the repository of all life information and contains both the blueprint for the construction of each part of every living organism and the recipe for how to bring the blueprint to life."

Q: "And, you have explained that this information is contained in code. Could you please refresh us about how that coded information comes about and works?"

A: "The DNA molecule, in coordination with the RNA molecule provides a complete encoding and decoding information system. The DNA information molecule represents an entire living organism. DNA is a coded language.

The DNA genetic code language uses an alphabet of four letters. Each of the letters represents a specific nucleotide nitrogenous base: T (thymine), A (adenine), C (cytosine), and G (guanine). The nitrogenous base letters are grouped into words of three letters each. The four nucleotide bases combine into three-letter sequences to form a word that describes one of 20 specific amino acids. For example, ACT*GTC*CAG* represents three coded genetic words (i.e., specific amino acids) strung together. Each three-letter word describes a specific amino acid. The exact sequencing of each of the three-letter nucleotide base groupings (called codons) is required in order for a specific protein to be built at the ribosome construction site in the cell. For most proteins the amino acids selected form a linear chain between 100 and 500 amino acids long, but some can number as great as 5,000."

Q: "You have explained that the simplest life found on Earth is in the form of bacteria that contains DNA. And, that bacteria contained coded information within segments of DNA called genes. And that at least 200 genes, coding for the construction of at least that many proteins, were required for life as we now know it to occur on Earth. Is that correct?"

A: "Yes."

Genetic Information Increases by Mistakes

Q: "How did evolution proceed to increase the information content contained in those 200 genes into the information content of our human genome containing over 30,000 genes?"

A: "Evolution by natural selection is the mechanism whereby the information content of DNA is increased. May I read a passage from our 2008 book *Science, Evolution and Creationism* which provides a very clear and concise explanation?"

Q: "Please do."

A: "Here is the passage:

'Contrary to a widespread public impression, biological evolution is not random, even though the biological changes that provide the raw material for evolution are not directed toward predetermined, specific goals. When DNA is being copied, mistakes in the copying process generate novel DNA sequences. These new sequences act as evolutionary 'experiments'.

Most mutations do not change traits or fitness. But some mutations give organisms traits that enhance their ability to survive and reproduce, while other mutations reduce the reproductive fitness of an organism.

The process by which organisms with advantageous variations have greater reproductive success over other organisms within a population is known as 'natural selection'. Over multiple generations, some populations of organisms subjected to natural selection may change in ways that make them better able to survive and reproduce in a given environment. Others may be unable to adapt to a changing environment and will become extinct.'

I hope that helps you Mr. Darrow."

Q: "So, the explanation is that while the biological process of evolution is not random the raw materials from which the process of natural selection selects are random. In essence, nature selects mistakes that improve things. Is that right?"

A: "That is correct. Those raw materials are mistakes in the copying process of DNA sequences. And, of course those DNA sequences we call genes. Natural selection selects from genetic mistakes."

Q: "Therefore, the ultimate source for increasing the information content of DNA in advanced organisms, including intelligent organisms like us, consists of mistakes or accidents in DNA? Is that your testimony?"

A: "Yes. As the ultimate source for increasing information content, that is correct."

Q: "What exactly is a mistake or accident in DNA?"

A: "A mistake or accident in DNA is quite simply a change in the nucleotide sequence in DNA. The simplest change occurs when one single nucleotide is exchanged for another. Since nucleotides serve to provide coded information, the information content is changed by that exchange. In order for the exchange to result in evolutionary consequences, the mistake or accident must occur in the DNA of the organism's germ line cells as opposed to its somatic cells."

Q: "Could you explain that a little further?"

A: "The normal bodily cells of plants and animals, indeed of all multi-celled organisms, are called somatic cells. All of the cells involved in the growth, repair and maintenance of the organism are called somatic cells. The special sex cells, or gametes, that are involved in sexual reproduction (sperm and egg) are called germ line cells. Only mutations in germ line cells will affect offspring. Thus, only germ line mutations will affect evolution by natural selection. As Carl Sagan put it in his book *Cosmos*:

> 'A mutation is a change in a nucleotide, copied in the next generation, which breeds true.'

Mr. Darrow, I hope you are starting to understand."

Q: "Do these mistakes in the DNA of germ line cells happen very often?"

A: "No, they don't. The mechanisms that reproduce genetic information in living organisms are incredibly efficient. And, when mistakes do occur, there are many enzymes that function to repair or replace damaged strands of DNA. The end result of the enzyme repair mechanisms is an error rate that is less than one mismatch in over one billion nucleotides. We have found that DNA replicates so accurately that it takes over five million replication generations to miscopy 1% of the language characters. But, in the long run, mistakes do happen."

Point Mutations

Q: "Are all genetic mutations limited to a single nucleotide change like Carl Sagan's book suggests?"

A: "No. Of course not. Sagan was simply making a generalization. He was referring to what we call 'point mutations'."

Q: "Would you please explain?"

A: "A mutation involving only a single base pair in a DNA molecule is called a point mutation. These point mutations are quite important in evolution, for evolution is basically a step-by-step process of minor changes that results in an evolutionary advantage. Usually a point mutation changes the sequence of a nucleotide resulting in a change of instructions for a single amino acid. However, a particular type of point mutation that is critical for evolution is what is known as a 'frameshift mutation' whereby a nucleotide is either added or deleted from a codon.

You will recall that codons, which provide the DNA instructions specifying a particular amino acid, are sequenced in groups of three non-overlapping DNA bases. In a frameshift mutation a single base is either added or removed from the sequence. That single mutation will then change all of the sequences coding for the amino acids following and thereby could provide instructions for constructing an entirely different protein by that single nucleotide change."

Q: "And, that one change can result in such a dramatic change accidentally?"

A: "Certainly."

Gene and Chromosome Duplications

Q: "Do other mutations occur beyond 'point mutations'?"

A: "Most certainly. The most likely major source of novel genetic material needed for evolution by natural selection is gene duplication. A gene duplication that results in faster evolutionary change is better-described as gene amplification, or better yet, chromosomal duplication. For a gene duplication event can result in a leap forward by the duplication of an entire gene or even many genes on a chromosome, or even an entire chromosome."

Q: "How does a gene duplication event happen?"

A: "The most common method of gene duplication results from what we call 'unequal crossing over' during meiosis. Meiosis, you will recall, is the reproductive process whereby single chromosome strands from the female and single chromosome strands from the male are united or 'cross over' to form a set of double-strand chromosomes containing base pairs of DNA. This 'crossing over' during meiosis occurs with great precision. However, very rarely, a misalignment occurs between a chromosome pair. This misalignment results in unequal pairing, know as 'unequal crossing-over'."

Q: "How does that affect the evolution of an increase in the information content in DNA?"

A: "Unequal crossing-over during meiosis results in one chromosome having two copies of a gene and the other chromosome deleting the gene entirely. The chromosome that now has two copies of a gene is in a position to advance evolution by natural selection better than a chromosome with only one copy of a gene.

   The two genes that exist after this type of gene duplication are called 'paralogs' and they evolve to code for proteins with different functions in the organism. The second copy of the gene is freed from the old function for it is not in play. It is like a second string player sitting on the bench. Because it is not in the game, it has no function and can mutate faster than a functional single-copy gene over many generations.

When mistakes are made, the natural DNA repair mechanisms will not be as concerned with the second-stringer. Once you have duplicate copies of a gene, mistakes that then occur during DNA replication can more quickly lead to the existence of modified proteins with new properties. Those modified proteins are the result of an increase in the information content of the organism's genome. Those modified proteins that prove advantageous for the species' survival are selected by natural selection and retained in the gene pool of the species."

Q: "Do these gene duplication events happen very often?"

A: "No, they don't. But they are very important. For example, humans and chimpanzees diverged from a common ancestor about six million years ago. During the ensuing six million years our DNA research indicates that there have been about 1,500 gene duplications in the total human genome of some 30,000 genes. Remember that almost all genetic mutations have either a detrimental or neutral effect. Only a miniscule few will provide for evolutionary advance. So, considering that fact, I think you must agree that such a relatively small number of gene duplications has produced some quite large results."

Q: "There certainly is a big difference between the two species. It is hard to understand how **random** gene duplications would have the amazing result of such a large-scale increase of genetic information in the human genome. From what you have just said it seems to me that the scientific evidence concerning gene duplications shows that the duplications are harmful instead of helpful. Isn't that true?"

A: "No. That is not entirely true. Quite the contrary, major evolutionary advances are made through gene and chromosomal duplication events."

Duplication Mistakes Make Gene Families

Q: "Let me pursue the point a little further. Isn't it true that unequal crossover is the mechanism that is believed by science to be the major cause of color blindness in humans? And, is not the duplication of a large part of chromosome 15 in humans one of the causes of mental retardation? And, isn't Downs Syndrome caused by an extra copy of Chromosome 21?"

A: "Those examples may be true. But they certainly are not the whole answer. Let me explain how unequal crossover serves to advance evolution by adding information content to DNA.

Science has classified gene sequences into families. Unequal crossover in genes and chromosomes gives rise to the evolution of gene families that have a common evolutionary origin but have evolved to perform different functions. The evidence for this is compelling. Through DNA analysis we trace back the historical development of different genes and confirm this fact again and again. Susumo Ohno in his book *Evolution by Gene Duplication* showed the way.

The primary evidence that duplication played the predominant role in adaptive evolution is the existence of these gene families. Members of a gene family share a common ancestor as the result of a gene duplication event. This gene family is found clustered within an organism's genome. Finding the same gene arrangement on the chromosomes of different species suggests common ancestry."

Q: "How exactly do such gene families develop through gene duplication?"

A: "Gene families consist of genes that encode proteins through similar, related structures but who have distinct functions. The explanation for each gene family is that multiple gene duplications, followed by random mutation and natural selection, produced a large family of genes with distinct functions that represent an increase in genetic information.

Let me give an example. The gene for a primordial oxygen-carrying protein duplicated. The non-functioning extra copy then mutated to a new gene encoding myoglobin (the oxygen-carrying protein of muscle). Then a further gene-duplication

unequal-crossover event resulted in another new gene that
encoded hemoglobin (the oxygen-carrying protein of red
blood cells). Then the hemoglobin gene duplicated and the
copies differentiated into the forms called alpha and beta.
Then those hemoglobin alpha and beta genes duplicated
several more times to produce hemoglobin alpha clusters and
hemoglobin beta clusters, like we see in modern vertebrates.

In this manner a large part of the increase in information
for modern genomes occurred through gene duplication
followed by natural selection of the random mutations of
duplicated genes that provided novel functions."

Q: "What is the scientific evidence in support of that evolutionary
explanation?"

A: "Quite simply, it is the existence of those gene families
themselves. The proof is that they exist."

The Randomness Explanation Requires Really Big Mistakes

Q: "Because you classify them in that manner does not provide
any evidence at all for the explanation that they came about
randomly. Does the existence of these so-called gene families
provide the answer for how all of the increase in information
content in DNA evolved?"

A: "No it does not. Science doesn't deal in absolutes. We are
always looking for more detailed explanations. The fact is that
the existence of gene families does not provide all the evidence
necessary to confirm the continuity of genetic development
that we are looking for. There is an important missing piece
of evidence."

Q: "What is that?"

A: "We have a very clear picture of the process and progress of
evolutionary development. All the facts fit together quite
nicely, except for the fact that from what we have discovered

to date about the nature of genetic duplications there was not enough time for all of the duplications to result in the products that we observe today. So, the missing piece of evidence is that which will support the 2R hypothesis."

Q: "What is the 2R hypothesis?"

A: "The large size and diversity of the vertebrate genome cannot be explained by individual gene duplication or even duplication events involving entire chromosomes. To resolve that discrepancy Susumu Ohno proposed that **two rounds of whole genome duplication** occurred at some point early on in the evolution of vertebrates, about 500 million years ago. That is the 2R hypothesis.

It is a scientific certainty that two complete genome duplication events would have provided a combination of possibilities that could have permitted a larger leap in evolution than single gene or even single chromosome duplication events could have."

Q: "Professor Tall, that explanation, again, includes a lot of 'would haves' and 'could haves'. Is there any scientific evidence in support of this 2R hypothesis?"

A: "The evidence comes from scientific studies that explore the relative positions of paralogs in the human genome. When looking at a subset of over 3,500 early vertebrate duplications, evidence was found that the duplications occurred in large segments. The collinear arrangement of these genes was mostly in a four-fold pattern. Since this repetitive pattern is seen across almost all the human chromosomes, it is unlikely that any combination of smaller independent duplication events could have generated the same pattern. That is the evidence."

Q: "So, the evidence that **proves** that two rounds of whole genome duplication events occurred early on in the evolution of vertebrates is that those two rounds of whole genome duplication events **just had to** occur in order to support what you observe today?"

A: "Mr. Darrow, again, that is a very sarcastic way to put it. I wish you could just understand science."

Q: "Professor Tall, believe me, I am trying."

A: "Yes, Mr. Darrow, you certainly are trying. Very trying."

* * *

Clarence Darrow again observed Goody Spyer vigorously at work with his comb. He then asked for and was granted a recess. Goody observed that all of the detailed material was wearing heavily on everyone in the courtroom. Both he and Judge Raulston were beginning to nod off. Goody advised the Plaintiff's lawyer that he needed to knock off the technical explanations and try to make things simpler. After the recess concluded, Clarence Darrow continued.

Mistakes in the Laboratory Never Improve Anything

Q: "What real evidence supporting random mutations has derived from real world laboratory experiments? Haven't all the lab experiments involving fruit flies, pardon the expression, flown in the face of increasing DNA information content through random mutation?"

A: "That is the standard canard used by the creationists. The fruit fly experiments were never aimed at proving that random mutations could increase information content. Radiation and poisons were used to mutate fruit flies in an effort to deform them so that the information gained by science could be used to benefit society."

Q: "In 1906 Thomas Hunt Morgan began to experiment with fruit flies in what was dubbed the 'fly room'. The specific species was *Drosophila melanogaster*. He selected that particular fruit fly for his genetic research because it was a fast breeder, producing a new generation every 12 days, and had only 4 chromosomes. These fruit flies have been subjected to x-rays and chemicals of all sorts for almost a century now. Isn't it true that over 3,000 different mutations of this fruit fly have been documented and that virtually none have proved beneficial by increasing the information content of the genome?"

A: "That may be true, but I would add, so what. As I said, the scientific experiments were never intended to show beneficial results from mutations."

Q: "A Russian geneticist, Theodosius Dobzhansky included the observation in his book *Genetics and the Origin of Species* that:

> 'The process of mutation is the only source of genetic variability and hence of evolution.'

Yet from his personal observations in Thomas Hunt Morgan's 'fly room' he concluded in his later book, *Evolution, Genetics, and Man*:

> 'Most mutants which arise in any organism are more or less disadvantageous to their possessors. The classical mutants obtained in *Drosophila* usually show deterioration, breakdown, or disappearance of some organs. Mutants are known which diminish the quantity or destroy the pigment in the eyes, and in the body reduce the wings, eyes, bristles, legs. Many mutants are, in fact lethal to their possessors. Mutants which equal the normal fly in vigor are a minority, and mutants that would make a major improvement of the normal organization in the normal environments are unknown.'

Dr. Tall, is it not a scientific fact that almost all mutations, whether produced by radiation or chemicals or environmental poisons or simply DNA replication errors, are either neutral or detrimental to the organism?"

A: "Yes. That is true. But, again, so what. Random mutation of genetic DNA is the only mechanism known to science from which natural selection can select which leads to an increase of information content in DNA. It is of no real consequence that such evolutionary successful random mutations rarely occur. The fact is that they do occur and that is all the proof we need."

Q: "Again, Professor, how do you know that such mutation is in fact **random** and undirected?"

A: "Quite obviously, since there is no known natural law that can account for genetic rearrangements, randomness is the only mechanism that comports with science."

Transposons Suggest Natural Genetic Engineeering

Q: "Okay, Dr. Tall. I don't want to beat a dead horse. Let's move on to wrap-up types of genetic mutations. We have covered point mutations, including frame-shift mutations. And, we have covered mutations by gene and chromosomal duplications, including the 2R hypothesis. Are there any other types of genetic mutations that science has discovered?"

A: "Yes. There are."

Q: "Would you please go on?"

A: "Yes, but I really would appreciate it if you would refrain from your sarcasm.

A remarkable class of duplications has been discovered wherein the duplicated region has popped up far away from its home site. This fascinating mechanism that can result in genetic change is what we call transposons."

Q: "What are transposons?"

A: "Transposons are mobile genetic elements that can change their positions within a genome. Some of these transpositions occur by a 'cut and paste' procedure while others are 'copy and paste'. Both procedures result in duplications of either single genes or long stretches of DNA containing numerous genes.

Some transposons duplicate via an RNA intermediate, known as retrotransposons. Indeed, such retrotransposons actually account for about 45% of the human genome."

Q: "Haven't some quite reputable scientists offered another explanation other than random mutations for the occurrence of these mobile genetic elements?"

A: "I am not aware of such an explanation."

Q: "Well, Sir, let me read an excerpt from a 2003 article in the *Boston Review* which was written by James Shapiro, Professor of Microbiology at the University of Chicago:

'. . . cells have molecular computing networks which process information about internal operations and about the external environment to make decisions controlling growth, movement and differentiation . . . . One can characterize this surveillance/inducible repair/checkpoint system as a molecular computational network demonstrating biologically useful properties of self-awareness and decision-making . . . . We are learning that virtually every aspect of cellular function is influenced by chemical messages detected, transmitted and interpreted by molecular relays. To a remarkable extent, therefore, contemporary biology has become a science of sensitivity, inter-and intra-cellular communication and control. Given the enormous complexity of living cells and the need to coordinate literally millions of biochemical events, it would be surprising if powerful cellular capacities for information processing did not manifest themselves.'

Dr. Tall, doesn't the explanation that living cells possess a feature akin to **natural genetic engineering** seem to you to be a much more plausible explanation than **random** mutations of DNA?"

A: "Most certainly not. There is no scientific evidence whatsoever to support the hypothesis that cells act in that fashion. They may seem to, but they do not. Since there is no known natural law that can otherwise account for genetic rearrangements, randomness is the only mechanism that comports with science."

Gene Complexity More Important than Gene Numbers

Q: "Again, nothing to be gained by beating that dead horse. I'll move on.
   But, just so I clearly understand. You believe that the principle method that results in evolution by natural selection to be that of gene duplication events. Is that correct?"

A: "That is correct. Gene duplication is the primary method."

Q: "And, the result of gene duplication is the evolution of increased information content of DNA?"

A: "Yes."

Q: "If evolution results primarily through gene duplication, then DNA information content should increase as the number of genes increases. Isn't that correct?"

A: "Yes it is. Duplicate genes evolve the existence of modified proteins exhibiting new properties. Those modified proteins that prove advantageous for species' survival are selected by natural selection and are retained in the gene pool. That is the explanation for how the information content of DNA increases by the development of new genes."

Q: "Professor, I find one particular aspect of that explanation to be most troubling. Isn't it true that science has discovered that simple organisms have DNA content and gene numbers comparable to that of advanced species? It has been discovered that a single-celled algae, Euglena, has a larger genome than a chicken. And, don't the genomes of both a corn plant and a rice plant actually contain more genes than the 30,000 genes contained in the human genome?"

A: "Yes. Those are accurate discoveries of science."

Q: "If these simpler organisms like corn and rice have developed more genes through gene duplication than we have, how can gene duplication be the principle method of evolution?"

A: "As usual, Mr. Darrow, you will try to twist things around. While the genomes of these plants technically contain more genes than the human genome they do not contain the more sophisticated genes that we possess. Our genome, while containing fewer genes contains more gene regulatory pathways with multiple levels of interactions whereby upstream transcription factors regulate downstream transcription factors, and so forth."

Q: "Then isn't it true, Professor, that not simple genetic duplication but, rather, the existence of exquisite inter-genetic regulatory sequences and gene regulation hierarchies provide the real determinates of an organism's complexity and DNA information content? And, isn't it true that no one knows how transcription factors actually activate these regulatory sequences and hierarchies?"

A: "That may be true. But, again, so what. All of the sequences and hierarchies and transcription factors themselves originated by the process of random genetic mutations. Science deals only with natural processes and the only natural process is randomness."

Q: "Again, I will lay off that poor dead horse. But, I must ask. Can you provide any examples at all of **random** genetic mutations increasing the information content of DNA?"

A: "Of course I can. Two pointed examples come to mind:
   • antibiotic resistance in bacteria; and
   • the human immune system."

Bacteria Solve Problems by Making Quick Mistakes

Q: "Would you please enlighten us with the scientific explanation for how bacteria acquire resistance to an antibiotic?"

A: "Certainly. Scientists have repeatedly performed experiments in the laboratory that prove beyond doubt that evolution by random mutation occurs to provide an evolutionary advantage in bacteria.

From a single solitary bacterium a population of succeeding generations is reproduced in the presence of an antibiotic. Many offspring die from exposure to the antibiotic. The offspring that survive are observed to have mutated genes that confer antibiotic resistance. That is a pointed example of increasing the information content of DNA through random mutation."

Q: "How exactly are the genes modified?"

A: "Basically, the mutated offspring that have developed resistance to the antibiotic have repressor genes disabled by the presence of certain enzymes in the bacterium. Those offspring whose repressor genes are disabled survive. Most pathogenic bacteria have succeeded in accumulating several sets of genes that provide them with resistance to a large number of antibiotics. This is generally done by degrading the specificity of a binding enzyme."

Q: "But, isn't the disabling of a repressor gene or the degrading of a binding enzyme actually an example of a **loss** of information, not a gain?"

A: "Mr. Darrow, you ask for an example, I give you a very clear one, and then you nit pick. The scientific truth is that resistance to antibiotics evolves by offspring inheriting a helpful gene that has randomly mutated."

Q: "Again, what evidence is there that the mutation was **random**?"

A: "And, yet again, the evidence is the fact that randomness is the only explanation that comports with natural processes and natural processes are the only possible scientific explanation."

The Immune System Protects Us by Making Quick Mistakes

Q: "The other example that you mentioned of how random genetic mutations increase the genetic information in DNA is the human immune system. Could you elaborate on that example?"

A: "I certainly could. And, I will. And, this one is clearly an example of an **increase** in DNA information content. No repressor genes are involved at all. Give me a few moments and I will explain."

* * *

Noah Tall took several minutes to locate the precise reference he needed to make his point. He then continued.

* * *

"An antigen (also known as a pathogen) is an invader like a virus or a bacteria or a parasite that is out to kill us. The human body has evolved a defense system that is called the acquired immune system. The acquired immune system is centered on what are called antibodies, large Y-shaped proteins that circulate in the bloodstream and which have evolved the ability to recognize and kill invaders. What we call B cells are white blood cell lymphocytes that produce the antibodies that kill the invaders. These B cells express unique receptors that cause antibodies to bind to a unique antigen and, thereby, then proceed to neutralize and kill a specific antigen.

Now this is how antibodies do their work. When the B cells encounter a foreign antigen invader that has never been encountered in the organism before, they engage a process that is known as hyper-mutation. The B cells quickly divide and begin to mutate at a very accelerated rate in order to develop an antibody that can tightly bind and thereby eliminate the antigen.

Since the antigen has never before been encountered, hyper-mutation in the B cells establishes a trial by error process. Through this process all possible mutations for the immune system are tried until the one that actually provides the information that is required to produce an antibody which is a match for and can destroy the new antigen is found. The modified B cell will then be selected to differentiate into long-lived plasma cells producing the specific antibody necessary to kill the new invader. After killing the invading antigen, it then produces memory cells that will serve to quickly oust the same antigen if it is encountered in the future through reinfection. That is how immunity is 'acquired'. This amazing process is sometimes called 'evolution in miniature' because it is a prime example of random mutation increasing genetic information content.

Through this hyper-mutation process the specific information required to disable the newly encountered antigen is successfully produced and retained by the organism."

Q: "How fast do these somatic hyper-mutations occur?"

A: "Extremely fast in comparison to normal germ line mutations. The order of magnitude is over a million times faster than germ line mutations."

Q: "But, then the antibody response through hyper-mutation is not really an example of evolution through natural selection necessary for evolutionary development is it? If the germ line mutations occurred by this method of hyper-mutation that rate of mutation would kill us wouldn't it?"

A: "That may be true, but that doesn't diminish the scientific fact that these mutations are random. They essentially quickly process all possibly solutions until the correct solution is found. It is like a five-number combination lock. If I had a lot of time I could try all the combinations manually until a solution is finally found that opens the lock. Of course, a computer could provide the solution and open the lock within a few seconds. That is a parallel to germ line mutation and hyper-mutation rates."

Q: "These hyper-mutations in the immune system seem more like computerized solutions to a problem than a random change of nucleotide bases. These hyper-mutations are certainly not random at all in the positions in DNA at which they occur, are they?"

A: "No. Hyper-mutation in the immune system is restricted to only very specific genes that encode for antibodies. And, they occur only in the small region where the change is needed in order to protect us and they occur only when they are switched-on by the controlling mechanism of B cells.

But, again, so what. They represent a concrete example of an increase in DNA information content by random mutations."

Evidence of Both Fast and Slow Genetic Change

Q: "In the NAS book *Science, Evolution and Creationism* you include this:

'The DNA evidence suggests that the basic mechanisms controlling biological form became established before or during the evolution of multicellular organisms and have been conserved with little modification ever since.'

It would seem that the take-home message from that evidence is that a lot of the fundamentals of biological structure have not been affected by DNA mutation. Is that correct?"

A: "Yes. That is correct. Many of the very early biological forms have been highly conserved."

Q: "In the NAS book *Science and Creationism* you explain:

> 'Because of mutations, the sequence of nucleotides in a gene gradually changes over time . . . .
>
> Genes evolve at different rates because, although mutation is a random event, some proteins are much more tolerant of changes in their amino acid sequence than are other proteins. For this reason, the genes that encode these more tolerant, less constrained proteins evolve faster. The average rate at which a particular kind of gene or protein evolves gives rise to the concept of a "molecular clock". Molecular clocks run rapidly for less constrained proteins and slowly for more constrained proteins, though they all time the same evolutionary event . . . .
>
> The clock for fibrinopeptides runs rapidly; 1 percent of the amino acids change in a little longer than 1 million years. At the other extreme, the molecular clock runs slowly for cytochrome c; a 1 percent change in amino acid sequence requires 20 million years. The hemoglobin clock is intermediate.'

So, science has discovered that when DNA mutations do affect biological form and structure that the pace of adaptive change through DNA mutations is pretty slow. Is that an accurate statement?"

A: "Yes. That is one of the points we were making by including that explanation in our book."

Q: "Yet, in the same two books you provide pointed examples of how evolution by natural selection by the same mechanism of DNA mutation occurs quite rapidly.

   In *Science and Creationism* you include this descriptor of fast adaptive change:

> 'The annual changes in influenza viruses and the emergence of antibiotic-resistant bacteria are both products of evolutionary forces. Indeed, the rapidity with which organisms with short generation times, such as bacteria and viruses, can evolve under the influence of their environments is of great medical significance. Many laboratory experiments have shown that, because of mutation and natural selection, such microorganisms can change in specific ways from those of immediately preceding generations.'

   In *Science, Evolution and Creationism* you include this descriptor:

> 'Over periods of just a few generations (or, in some documented cases, even a single generation), evolution produces relatively small-scale microevolutionary changes in organisms. For example, many disease-causing bacteria have been evolving increased resistance to antibiotics. When a bacterium undergoes a genetic change that increases its ability to resist the effects of an antibiotic, that bacterium can survive and produce more copies of itself while nonresistant bacteria are being killed. Bacteria that cause tuberculosis, meningitis, staph infections, sexually transmitted diseases, and other illnesses have all become serious problems as they have developed resistance to an increasing number of antibiotics.'

Professor Tall, this is an important point that I want to clearly understand. Is it your testimony that the same mechanism of **random** mutation of genetic DNA explains adaptive change that takes place over millions and millions of years and also over the course of a single generation?"

A: "That is correct. Both are the product of randomness."

Q: "You explain on one hand that science has discovered the rate of amino acid realignments takes millions of years. One percent, just one percent, realignment takes between one and twenty million years. And you explain that you are quite sure that the change agent was **random** mutation.

Then you explain that tiny bacteria can change their DNA in the course of a single generation in such a completely adaptive way that they can avoid being killed by antibiotic serums that were developed by brilliant, very intentional, scientists in a well-thought-out plan to kill them. In essence, bacteria have no problem in quickly changing their amino acid sequences to avoid being killed. And, the change agent that they employ to out-smart brilliant scientists is **random** mutation of their genetic DNA.

Dr. Tall, how does that explanation possibly make sense to you?"

A: "There you go trying to twist things around again. Mr. Darrow if you want to believe that there is some unseen intelligent agent that causes these things that is fine with me. That is simply not science. The randomness explanation for both slow and fast genetic change is the only explanation possible if we are to remain scientific. I don't know how many times I have to explain that."

Q: "Dr. Tall, you have indeed provided that same explanation many times. And, you have yet to provide any confirmable scientific evidence at all to support the inference that all DNA mutations are the result of randomness.

You have now testified that the only possible mechanism from which natural selection can select for all adaptive trait modifications in living organisms is random mutations of DNA. You have testified that scientific evidence shows that DNA replicates so accurately that it takes over five million replication generations to miscopy 1% of the language characters. You have also testified that the most basic living organisms – single celled bacteria – can evolve resistance to antibiotics within a single generation.

In essence, tiny bacteria and viruses routinely outsmart our most brilliant scientists who are developing antibiotics to protect us from infection and disease. Some of these scientists are now engaged in gene splicing to develop new cures. If scientists are successful in their gene splicing endeavors they are rightfully lauded for their brilliant work. Yet, it is simply no big deal to scientists when they observe single-celled creatures splicing their genes to provide helpful modifications routinely. You devoutly maintain that the only mechanism that produces this adaptive change is **randomness**.

How can you continue to provide the same pat explanation for such diverse and wondrous things?"

A: "I have given you my answer many times, Mr. Darrow. You simply cannot or will not understand."

* * *

Clarence Darrow paused in his questioning as he saw Goody Spyer arranging his comb-over with vigor. He asked Judge Raulston for a recess, which was granted, and conferred with Goody.

Goody was very blunt with his insistence that the Judge was getting a glazed-over look with all of the detailed questioning. Clarence Darrow had made his point time and time again. It was time to wrap things up. Not only the Judge, but Goody as well, would soon tune out entirely unless Clarence Darrow could bring some focus again to the main point of the trial.

When Judge Raulston reconvened the session Clarence Darrow heeded Goody's advice.

* * *

Q: "Professor Tall, I've about reached the end. Your educational sessions on what you believe should be taught in a public school biology class have been very enlightening.

From my notes of your testimony I would like to summarize what I have learned from you concerning the things that should be included in public school science class. I would ask that you simply provide a yes or no answer, without elaboration to the following questions. Will you do that?"

A: "I would be happy to. However, if I find you to be misleading in your questions I will have to elaborate."

Q: "Okay. First, did life on Earth originate from natural causes accidentally and without direction?"

A: "Yes."

Q: "Was the mechanism for the origination of the first living organisms on Earth the natural selection of random mutations of non-living molecules that provided an evolutionary advantage?"

A: "Yes. That is basically true."

Q: "So, non-living organic molecules evolved by the same process as living organisms, through the process of modifying genetic information, even before genes existed?"

A: "See, you try to twist things. Of course genetic change could not occur before genes existed. But, yes, the non-living molecules that evolved into genes did undergo the same type of evolutionary process of natural selection through random mutation."

Q: "Sorry. I really do not intend to be misleading. It is important that we clearly understand your explanation. I'll continue.

Did the information originally contained in DNA originate from natural causes accidentally and without direction, by the natural selection of random mutations of **something** that provided an evolutionary advantage?"

A: "Yes."

Q: "So, the coded language structure of DNA was the product of natural causes with no direction or intention or purpose involved?"

A: "That is true. I have told you that many times. There was no intention or direction or purpose involved, and nothing supernatural was necessary."

Q: "You have explained that in order for the information in DNA to be of use to a living organism the information must first be organized to do at least three distinct things:
   • It must be transcribed into RNA format.
   • It must be transferred to the site in the cell where proteins are to be constructed.
   • It must be translated from the DNA / RNA coded language into the amino acid language necessary for the construction of proteins through a long chain of different amino acids.

You would teach that these three distinct steps evolved through the process of the natural selection of random mutations of genetic DNA that served to provide a survival advantage for the living organism. Is that the correct explanation?"

A: "No. Actually I would teach that evolution by natural selection of random genetic mutations provides a survival advantage to the **offspring** of the living organism."

Q: "I stand corrected. But this three-step mechanism, whereby protein construction evolved, was a natural process, without intention or direction or purpose. Is that correct?"

A: "Yes. That is correct."

Q: "You have explained that single-celled bacteria evolved the ability to obtain the energy they needed for life directly from sunshine through the elegant process of photosynthesis. The NAS explanation is that these tiny bacteria did that by accident, without direction or intention or purpose. Yet, all of the purposeful efforts of our brilliant scientists today have not been able to duplicate that process. That's correct, is it not?"

A: "Yes. But, one day we will accomplish that."

Q: "Those bacteria and early plants evolved the process of photosynthesis. But then, in later evolutionary development, the early animals, by random mutations of genetic DNA, un-evolved that process and evolved the process of eating the plant synthesizers and gaining their energy by the complicated processes of glycolosis and cellular respiration. And, both of those processes are catalyzed by numerous enzymes that evolved in order to control the release of energy in a step-by-step fashion so that animal organisms don't just spontaneously combust. All of this was done by random mutations of genetic DNA without intention or direction or purpose. Is that correct?"

A: "Despite your dripping sarcasm, the answer is yes. Those developments were quite natural, without intention or direction or purpose. They fit very nicely with the fact that photosynthesis had drastically altered the atmosphere of the planet and that plant-eating living organisms were needed to balance things out in that new environment.   But, to correct the record, I never said **un**-evolved. That's your term."

Q: "Then, the early animals evolved the process of activating and maintaining a complex electrical circuitry in each and every one of those living animal organisms.   And, the mechanism by which electrical circuitry evolved was the natural selection of random mutations of genetic DNA without intention or direction or purpose. Right?"

A: "Right again. This is sophomoric and tiresome, but the answer is yes."

Q: "And, to maintain that complex electrical circuitry, these living organisms developed and maintained the operation of sodium-potassium pumps within each cell. These pumps are absolutely essential to the maintenance of electrical charge, and animal electrical circuitry cannot continue to function without these pumps. About one-third of the energy that animals consume is dedicated to the operation of these pumps which actually reverse a basic law of physics, the equivalent to making water run uphill. Is that correct."

A: "That's redactive.   But essentially yes."

Q: "They accomplished this feat by the random mutations of genetic DNA without intention or direction or purpose?"

A: "Yet again, that is correct. Evolution of each and every aspect of life developed by the mechanism of randomness. That is the only explanation that is scientific."

Q: "Then, over the course of the past 500 million years or so the same mechanism of random mutations of genetic DNA produced more advanced animal life that evolved to include hearts and lungs and kidneys and brains and blood clotting and immune systems?"

A: "Yes. Mr. Darrow we have been over all of this before. The only mechanism that is possible for science is randomness. That will always be the answer."

* * *

Clarence Darrow walked to the lectern and scanned some papers he had earlier placed there before continuing. As Goody Spyer had told him during the last recess, he needed to finish his case by stressing the extent of the impact of the NAS teachings. And, Clarence Darrow could only do that through the testimony of Noah Tall. He then continued.

* * *

Q: "Professor Tall, you have just testified at great length in this trial that the randomness explanation is, at core, the only possible mechanism for biological change. You have confirmed that the randomness explanation is the heart of the two NAS publications that the Dayton School District requires Mr. Scopes to use in his biology class. In short, those books maintain that random adaptive mutations of non-living molecules resulted in the origin of life on Earth. Further, random mutations of genetic DNA resulted in the adaptive evolution of all living organisms that have ever resided on this planet. It is important that I do not misconstrue your testimony. Is the recap that I have just given correct?"

A: "Yes. No matter how often you ask the question the answer will always remain the same. The randomness explanation is the only possible explanation if we are to remain scientific."

Q: "Is the extent of that NAS teaching limited to just the students in the Dayton School District?"

A: "Of course not."

Q: "Indeed. That NAS teaching is in fact very widespread. That teaching is widely included in textbooks used in public school biology classes and is included in instructional materials made available on the Internet for free to public school science teachers.

A most significant website that I will use as an example is maintained by Think Quest. Thinkquest.org is a website established by the Oracle Education Foundation. It publishes an online learning platform designed to enable teachers to integrate learning projects into their classrooms. Think Quest also provides a professional development program for educators. These services are provided for free to some 400,000 participants. Are you familiar with these services?"

A: "I most certainly am. The Oracle Education Foundation is a true asset for science education in this country."

Q: "I'd like to review some of the materials on the subject of evolution that are provided by the Oracle Education Foundation through Think Quest and see if they accurately reflect the teaching of the NAS. Will you help me with that?"

A: "I certainly will. I can't imagine that you will find any surprises there."

Q: "In discussing the subject of natural selection Think Quest provides the following explanation:

'Therefore, natural selection is dependent on the existence of mutations in the genes coding for different characteristics of an organism. Most mutations in DNA are spontaneous and random, sometimes caused by passing cosmic rays or other exposure to radiation. Mutations may also be caused by errors in the formation of the genes in the parents' gametes . . . .

A vast majority of mutations in an organism's DNA have deleterious effects on the organism and thus will be immediately selected against, or they will be irrelevant or have only very marginal effects. Only a tiny percentage of all mutations will confer a survival advantage on the organism that inherits it . . . .

Natural selection is quick to seize upon the very rare beneficial mutations that arise. Even a small survival advantage will be selected for over generations, eventually saturating the overall gene pool with the altered gene. Natural selection may depend on random mutations, and its operation may be slow and fitful, but it is extremely efficient in "weeding out" successful adaptations to be passed on to future generations.'

Professor Tall, does that explanation comport with the NAS explanation?"

A: "Of course it does. The Oracle Education Foundation is committed to scientific explanations."

Q: "In addressing the subject of the origin of life, Think Quest explains that the Primordial Soup Theory, first suggested by Charles Darwin:

'. . . states that self-replication entities, the precursors to life as we know it, arose spontaneously out of the chemical environment of the early Earth. This theory argues that the chance reactions taking place at high rates in the chemical mixture of the early atmosphere eventually gave rise to molecules with the property of replication . . . .

At first, the new replicatior would have free run of the chemical resources in the primordial soup. The replicators would probably reproduce freely under such conditions because they would have a monopoly on the available resources. In addition, the sheer numbers of copies being made, added to the fact that the replicator would be very primitive and without editing mechanisms, would result in numerous copying errors.

These errors are mutations that will later be used in the development of natural selection . . . . '

Dr. Tall, doesn't that explanation by Think Quest comport completely with this 'scientific truth' that the Dayton School District requires Mr. Scopes to use in his biology class that is contained in the 2nd edition of *Science and Creationism: A View from the National Academy of Sciences*?

'For those who are studying the origin of life, the question is no longer whether life could have originated by chemical processes involving nonbiological components. The question instead has become which of many pathways might have been followed to produce the first cells.'

Doesn't the Think Quest explanation simply mirror the NAS explanation?"

A: "Of course it does. Again, Mr. Darrow, the Oracle Education Foundation is committed to scientific explanations."

Q: "You teach that the randomness explanation for the conception and evolution of all life on Earth is a 'scientific truth'. That teaching actually serves to shut the door on a great deal of scientific exploration. And, it certainly has a devastating effect on society.

Any thinking youngster in science class cannot possibly reconcile this 'scientific truth' that all living things developed by accident, without direction or purpose or meaning, with a societal idea that there is actual purpose and meaning in life. For that youngster you have closed the door to the idea that there is any real purpose and meaning in life beyond that which we simply make up.

And, any thinking person cannot believe in the validity of any ethical system that is not based solely on situational ethics if this 'scientific truth' is in fact ultimate reality.

Professor Tall, don't you understand that teaching your belief that randomness is, in fact, ultimate reality has this devastating effect on society?"

A: "Mr. Darrow, that is not my concern. I am a scientist. I am concerned only with teaching the truth of science."

* * *

Clarence Darrow closed the notebook he had been using on the lectern before him. He smiled at Noah Tall and shook his head slowly side to side. He then proceeded to conclusion.

* * *

"With that, Professor Tall, I will stop and thank you for your patience and your truthfulness during these several trying days.

Your Honor, we have now established through Professor Tall's testimony that there is no scientific basis whatsoever for the randomness explanation provided in the NAS publications for the creation and evolution of life on Earth.

And, we have established through his testimony that this same randomness explanation is widely reiterated to science teachers and students throughout this country as 'scientific truth'. Most importantly, we have shown by Professor Tall's expert testimony that there is no scientific basis whatsoever to support the widespread teaching of this metaphysical doctrine of randomness that is the core belief of the religion of scientific atheism.

Your Honor, I have no further questions. The Plaintiff rests."

CHAPTER 10

The Evening of the Sixth Day

After Clarence Darrow rested the case for the Plaintiff, Judge Raulston adjourned the sixth day of the trial. Goody Spyer's consulting work was now done. He could give Clarence Darrow no further advice on how successfully he was making his case before the Judge. Closing arguments would provide no opportunity for courtroom consultations.

At the urging of his wife, Sarah, Goody invited me to dinner at their place. I hadn't had a home-cooked meal in over a week and I looked forward to it eagerly.

Goody and Sarah lived a good stretch outside of town where they had a 20-acre 'hobby farm'. They lived in a brick cottage farmhouse overlooking a two-acre pond that was stocked with catfish and black crappie. Since their retirement their greatest farming venture was the never-ending battle of defending their vegetable garden from the seeming hoards of marauding rabbits and deer that were determined to reap what Goody and Sarah had sown.

As I previously related, Goody was my basketball coach when I attended Hazel High School. Goody was a farmer, not a teacher at the school. You see, the good folks of Hazel had the common sense to recognize that teaching history or English or science is not a prerequisite for coaching high school basketball. Goody was selected to be the high school basketball coach for over thirty years even while keeping his day job as a farmer. He was chosen to be coach because he had been a star player when he was in high school and because he was really good at employing teamwork.

Goody and Sarah had sold their farm in western Kentucky and now enjoyed a comfortable retirement in Tennessee. The countryside and the weather were pretty much the same and they could now spend more time with their daughter Jennie and their two grandchildren who live in Dayton.

When I arrived for dinner Sarah greeted me with an enormous hug and insisted on showing me around the place before eating. My overall impression was a simple one. When I retire this is what I want. All I needed now was to get a hefty raise from Uncle Rob and to marry a sweetheart like Sarah.

I have often observed that ordinary folks, with an unexceptional formal education, have an innate ability to understand complex things in a straightforward manner. Clarence Darrow had recognized that quality in Goody and sought his advice. Conversation both during and after that evening's meal provided ample evidence to me that Sarah had the same ability.

During a dinner of chicken and biscuits and all the trimmings small talk was pretty much confined to catching up on personal things, what relatives are doing what, how they are faring healthwise. That sort of thing. But, as we enjoyed a dessert of apple pie and ice cream and then settled into easy chairs for coffee, Sarah broached the subject of the trial.

"You know, Shorty, this trial has been broadcast over the radio ever since the start and a lot of it has been reported on TV. I have been listening on the radio off and on over the last few days. And, in the evenings, I've been talking with Goody about how Mr. Darrow has been doing. I've got some opinions, but I'd like to know what you think about the whole thing Shorty."

"Well, Sarah, I'm covering the trial as a reporter. A news reporter is supposed to do just that. Report the news. I'm supposed to accurately report just what is happening. I'm not supposed to be biased one way or the other. I think that is important. So I don't want to really get into my personal opinions until after the trial is over and I have written my articles for the paper."

I took a few sips of coffee in hopes that she wouldn't press the point and then tossed the ball back to her. "But, I would really like to know your thoughts. Knowing what folks without a direct interest in the trial are thinking about it is itself news."

"Well, Shorty, Goody has always said that I can't hold my tongue even if I should. I'm not about to start now. So, here goes." Sarah daubed her mouth with a napkin and continued.

"After listening to all those exchanges about all that complicated stuff that Dr. Tall has explained I guess my feeling is simple. I just don't understand how someone as smart and as educated as Dr. Tall can be so stubborn. After you listen to all of the explanations the bottom line remains the same. He is convinced that accidents and mistakes make complicated things happen. He says that evolution by natural selection occurs when nature randomly selects from things that just happen by mistake or accident. That just doesn't make common sense. And, I think that Mr. Darrow has done a good job of pressing him for evidence to prove his point. Dr. Tall has never provided any real evidence for his explanation."

Goody Spyer smiled broadly at me as he recognized that his wife of nearly forty years was just getting warmed up. She continued.

"For instance, when they talked about how birds build their nests. It would probably help Dr. Tall a lot to actually observe a bird building a nest for her chicks. There is nothing accidental about it. Accidental changes in DNA did not result in making bird nests or in the parents sitting on the eggs till they hatch or in teaching the kids how to fly. All he would have to do is watch a goldfinch nest for a while. Observe the parents building the nest and then watch them coax the chicks into making their first flight. You might chalk all that up to instinct, but that did not happen by accident."

At the risk of being trampled I interjected a comment. "Sarah, don't you think that natural selection is a good explanation for a lot of what is observed?"

"Sure I do Shorty. The scientists have made it clear that there is a lot of evidence that we all evolved over time from simpler things to more complex things. It's just that going from simple to complex by accident doesn't make any sense. There would have to be a lot of scientific evidence to support that explanation, and from what I have heard, the scientists don't have any real evidence that all of evolution simply happened by accident.

Like that stuff about photosynthesis happening by accident. And then animals losing the ability of photosynthesis because losing that ability gave them an advantage for survival. I mean, get real.

To get the energy I need to live I sure have to expend a lot of energy in the first place. First, I've got to plant the cornfield. Then harvest the crop. Then feed the corn to the chicken. Then kill the chicken, clean the chicken, and do the cooking. Finally I've got to eat the chicken and the corn. Only then do I get the energy I need.

If the good doctor ever had to do all that then I think he may better understand that mistakes and accidents wouldn't discard such an elegant system for getting energy as photosynthesis in favor of the complicated one I have to use. It would sure be a lot simpler for me to just sit out in the sunshine and soak my energy in." At this point she opened her arms wide with a gesture to emphasize her point. But, she wasn't through yet.

"Noah Tall won a Nobel Prize in Medicine for developing a cure for rheumatoid arthritis, a crippling disease that plagued our daughter, Jennie, for years. Our family owes Dr. Tall a debt of gratitude that we can never repay. But, that does not mean that his belief that accidents and mistakes cause complicated things like photosynthesis and cellular electrical charge to happen makes any sense." Sarah poured another cup of coffee for me before continuing.

"There is harmony everywhere in nature. And, there is also savagery everywhere in nature. If you spend any time at all on the farm or in the woods you observe a lot of both. The butterflies and hummingbirds and bees pollinate the brightly colored flowers. The owl impales the helpless bunny rabbit with its talons, while the hognose snake swallows the unwary chipmunk whole and slowly kills him with his digestive juices.

No one knows why both harmony and savagery exist always side by side in nature, but they do. Dr. Tall's simple explanation is that random mutations of genetic DNA is the final answer for everything. That may be enough for him as a scientist. But it's not enough for us country people who have been paying attention to life and living."

Sarah gave out a long sigh as she contemplated the potential in this courtroom struggle that would surely be characterized again by the media as science pitted against religion. Her good country neighbors would almost ensure that. They seemed dedicated to the banishment of the blasphemous theory of evolution, no matter the evidence. She continued.

"Shorty, you have seen how a lot of the folks around here react when the big-wigs of science lecture us about how they just know that the Christian God doesn't exist. When they explain that nothing outside of nature is necessary to explain all that happens within nature they tell us that believing in God is really stupid. And that creates battlelines.

Pastor Andrews has been telling his congregation to attend the trial and he expects them to crowd the courtroom on the final day. He wants those who can't find a seat to picket outside the courthouse. The real religious zealots who preach that everything in the Bible should be taken literally are organizing protests that miss the real point of Mr. Scopes' lawsuit entirely. Posterboards that proclaim 'Evolution is Sin' and 'Darwin's God is the Devil' will likely work against the point of Mr. Scopes' lawsuit.

Mr. Scopes wants no religion to be taught in science class. The Bible-bangers want the Bible's creation story to be taught there. I'm sure that Judge Raulston will be outraged by the zealots' tactics and will reject their religious piety outright. But I don't know what effect, if any, their shenanigans will have on the Judge's ruling.

Rejecting religious piety is no reason for upholding scientific piety. Two wrongs never made a right. I think it would make a lot of sense for Judge Raulston to see that and keep the teaching of both traditional religion and atheism out of the classroom."

Sarah was finished. She abruptly ended the conversation (actually her monologue). She excused herself and cheerfully turned her attention to clearing the dining table and doing the dishes.

I thanked Goody and Sarah for their hospitality and returned to my hotel. Tomorrow would be a long day.

"We have no government armed with power capable of contending with human passions unbridled by morality and religion. Our Constitution was made only for a moral and religious people. It is wholly inadequate to the government of any other."

John Adams
Founding Father
Signer of the Bill of Rights
Second President of the United States

CHAPTER 11

The Trial – The Seventh Day
Closing Arguments

On the morning of the seventh day the courtroom was crowded. An overflow seating area had been set-up in the cafeteria, located in the basement of the courthouse, complete with a video and sound system so that all interested attendees could see and hear the closing arguments.

Inside the courtroom the bailiff conducted the perfunctory opening ritual, the Judge was seated at the bench and the audience and parties were seated. Judge Raulston nodded to Clarence Darrow, John Scopes' attorney, to proceed with his closing argument.

The Plaintiff's Closing Argument

"Your Honor, I begin by thanking you and Mr. Bryan and, especially, Dr. Noah Tall for the extraordinary patience and courtesy extended to me over a long period of tedious examination of many incredible things. The biology lesson provided by Professor Tall has been invaluable.

The legal system has had a long history of reviewing the controversial subject of evolution by natural selection. The celebrated case of what was termed the 'Monkey Trial' was conducted in this very town in 1925. At a time when broadcast radio was just beginning, that new information medium was used to provide a science lesson to the nation.

We have come a long way since then. High definition television now provides us with news on every subject imaginable 24 hours a day, 7 days a week. And, since 1925 dedicated scientists have made incredible discoveries about each and every aspect of biology.

Most of us have a great reverence for science and scientists. Little wonder in that. We believe in science for we see daily the benefits and the marvels that scientists have brought into our lives. Household computers, cell phones and plasma TVs have become part of our daily utensils. Surgeons are using lasers to remove tumors from our bodies and correct lifelong eyesight deficiencies to leave us free of eyeglasses forever. Organ transplants are everyday occurrences, wonder drugs ease our pain and suffering, DNA matches are used to convict criminals, and the human genome has been mapped. The scientists are the smart guys who are doing all these incredible things.

We see the truthful results of science each and every day of our lives. We trust science. Scientists teach their explanations based only on facts, don't they? If these really intelligent scientists have factually determined that every living thing on Earth is the result of mere chance, of the natural selection of **random** mutations of the information contained in DNA molecules, then that is just what we should teach our children in our public schools.

My client, Mr. John Scopes, is a scientist and a high school teacher. He does **not** believe that our most eminent scientists have factually determined that mutations of genetic DNA are **random** at all. He has looked at the evidence and concluded that it is wholly lacking to support the inference that **random** mutations of genetic DNA molecules provided the mechanism to produce all of the wonders of modern living organisms that we see today. He believes that the most elite scientists of this country are committed to the **belief** that the change mechanism of random mutations is in fact true. Where physical evidence ends, belief beyond the physical – **metaphysical belief** – begins.

Mr. Scopes, as a science teacher, strongly objects to such a metaphysical belief being taught as true in his science class.

The single issue before this court is whether the ultimate and overarching explanation of **randomness** is a scientific or a metaphysical-religious explanation. The Law in the State of Tennessee and the Constitution of the United States both require that only scientific explanations may be taught in science class.

We have spent the past six days exploring that very issue with a lot of specificity. It has not been a pleasant process but it has been a necessary one. Determining whether randomness is a scientific explanation or a metaphysical- religious explanation necessarily depends on the actual evidence discovered by science.

Generalized speculations are not the determinants. Facts are. So, in presenting my closing argument, I will proceed to review the most compelling facts obtained through Professor Tall's testimony. And, of course, we have stipulated that his testimony is representative of the position of the National Academy of Sciences."

* * *

Clarence Darrow paused for what seemed several minutes while he arranged his notes on the Plaintiff's table. He took a few sips from his water glass and then continued.

* * *

"The *'Only Science in Science Class Act'* provides a very clear standard for explanations that may be taught in science class and for those that are to be excluded. Only scientific explanations are permissible. And the elements necessary to qualify as a scientific explanation are very specific.
- First, only natural explanations are allowed.
- Second, the natural explanation must be inferred from confirmable data whose results can be observed, tested, replicated and verified.
- Third, inferences that cannot be inferred from confirmable data are deemed to be doctrine of a religion and cannot be taught.
- Fourth, and this one is a little statutorily redundant, explanations that cannot be based on empirical evidence are not a part of science.

It is interesting to note that this standard is, in essence, the same standard advocated by the National Academy of Sciences in their publications. I would imagine that the Tennessee legislature actually used the NAS language as a model. That would certainly make sense, for there is no more prestigious scientific organization in America than the NAS.

Now, let's recap the NAS explanations for the biological phenomena that we have covered in the last six days. Dr. Tall, the President of the NAS, has provided very pointed explanations concerning a wide array of amazing biological things. The explanation provided for each and every one of those phenomena, he contends, should be taught in science class. Indeed, many of those explanations are included in the two NAS books that the School District has told Mr. Scopes he must teach or else he will lose his job.

The ultimate and overarching explanation that the NAS and the School District want Mr. Scopes to teach in science class for each of these amazing biological things is that they occurred naturally through the process of evolution by natural selection of **random** mutations of genetic DNA. The NAS is adamant that science class must include the explanation that everything that lives on Earth is the product of purely natural causes. They insist that public school science teachers must instruct impressionable youngsters that science has discovered that life on Earth is ultimately the product of randomness, without intention, without direction, and without purpose.

Mr. Scopes believes that explanation to be unscientific, outside the scope of science, and should be excluded from his science class.

Let's turn to the evidence adduced through this trial and see what it has revealed."

Explanation #1

"The NAS explanation for the creation of life on Earth is that non-living molecules became more and more complex by the process of natural selection of random mutations of nonbiological molecules.

Scientific evidence reveals that the simplest living organism that ever lived must have had at least 200 genes coding for at least 200 different proteins. Scientific evidence reveals that at least this number of different proteins were required for an organism to be alive. The odds of constructing a single protein by random chance are astronomically greater than the odds of selecting one particular atom from among all the atoms in the entire universe. The odds of selecting all 200 proteins at the same time by chance are, of course, exponentially greater.

The scientific laws of statistical probability evidence that when you deal with such fantastic probabilities there reaches a point where the highly-improbable becomes impossible. In short, the laws of statistical probability reveal that the creation of life by random chance was and is **impossible**.

Scientific evidence tells us that the first life on Earth required an organism to do at least three discrete things:
- The living organism had to store and process information.
- The living organism had to acquire and use energy.
- The living organism had to reproduce its cells and itself.

Scientific evidence reveals that scientists conducting laboratory experiments for over the past fifty years have never been able to get non-living molecules to do any one of these three things, let alone all three things together. Scientists have never been able to create life from non-living molecules. Brilliant scientific minds, filled with direction and intention and purpose, have been wholly unsuccessful in creating life from nonbiological components in the laboratory. But, the NAS is quite certain that life was created quite naturally from non-living molecules through the natural selection of random mutations. One of the NAS books that the School Board requires Mr. Scopes to use in his biology class includes this pointed statement:

> 'For those studying the origin of life, the question is no longer whether life could have originated by chemical processes involving nonbiological components. The question instead has become which of many pathways might have been followed to produce the first cells.'

In essence, the NAS wants our public school students to be told in science class that there is no longer any doubt that life just happened, by accident, without any direction or intention or purpose.

There is no confirmable data whatsoever in support of that explanation. It is an inference based on no scientific evidence. It is an explanation spun from hole cloth.

Because that explanation is not based on empirical evidence it is deemed to be doctrine of a religion and cannot be included in science class. It is a metaphysical-religious explanation that should be expelled from science class."

* * *

Clarence Darrow organized his notes, took a sip of water and proceeded.

* * *

Explanation #2

"Let's turn now to the NAS explanation for how the first living organisms on Earth, single-celled bacteria, became more and more complex. The National Academy of Sciences explains the process as evolution by natural selection of **random** genetic mutations of DNA.

The first incredible phenomenon that we examined involving single-celled bacteria was photosynthesis. Photosynthesis is the process whereby single-celled bacteria stopped eating rocks and began to acquire the energy they needed for living by capturing the energy of sunshine and converting it into a safe and usable energy form that could be stored for future use.

The scientific evidence reveals that the process of photosynthesis is so exquisite and complex that today's scientists have been unable to create photosynthesis artificially in the laboratory. Brilliant scientific minds, filled with direction and intention and purpose, have been wholly unsuccessful in creating artificial photosynthesis in the laboratory. But, the NAS is quite certain that single-celled bacteria developed the process of photosynthesis quite naturally through evolution by natural selection of random mutations of genetic DNA. No scientific evidence is offered in support of the NAS explanation of randomness.

There is no confirmable data whatsoever in support of that explanation. It is an inference based on no scientific evidence. It is an explanation spun from hole cloth.

Because that explanation is not based on empirical evidence it is deemed to be doctrine of a religion and cannot be included in science class. It is a metaphysical-religious explanation that should be expelled from science class."

Explanations #3 and #4

"After bacteria and green plants evolved the exquisite process of photosynthesis more advanced organisms seem to have un-evolved the process. Animals cannot acquire the energy they need for life from the Sun. Animals must eat other living things or things that used to be alive in order to obtain the energy they need for life. As animals un-evolved photosynthesis they acquired the ability to establish and maintain a complex electrical circuitry within their bodies.

The scientific evidence shows that animals did this by constructing sodium-potassium pumps. And, scientific evidence shows that animals must expend over one-third of the energy they acquire from eating other living things in order to maintain these pumps. The scientific evidence shows that these sodium-potassium pumps are actual pumps that are constructed and operated in all living animal cells. They are absolutely essential for maintaining cellular electrical charge that animals require for life.

The NAS explanation is that evolution by natural selection of random mutations of genetic DNA results in adaptive change in living organisms that provides a solution to an environmental challenge. The adaptive change that provided a solution to an environmental challenge is retained in the species' gene pool.

There is no scientific evidence that explains the environmental challenge that would have caused random genetic mutations of DNA to rid the animal organism of the ability to gain its energy directly from sunshine. In fact, as animals **un-**evolved the ability to gain energy directly from sunshine they encountered enormous challenges that had to be overcome. The intricate and elegant solutions to those problems included:

- Development of a complex electrical circuitry necessary for sustaining animal life, along with a sophisticated pumping mechanism necessary for maintaining a constant electrical charge. Scientific evidence reveals that both of these changes occurred at about the same evolutionary time.
- Development of an energy chemical processing plant mediated by specialized protein enzymes that are needed in order to release the high-energy contained in glucose in a controlled step-by-step fashion necessary to prevent the living organism from being destroyed by spontaneous combustion.

The NAS explains that it is not important to be able to identify the specific environmental challenge that would have produced these amazing and unexpected results naturally. We are told that we do not need to know the specific challenge. The NAS explanation is that there **just had to be** an environmental challenge that produced these amazing results because that is what evolution by natural selection of random mutations of genetic DNA requires. And, the NAS explanations simply ignore the scientific evidence that:

- The construction and operation of actual pumps necessary to maintain cellular charge requires cells to construct a complex mechanism that actually reverses the natural forces of physics.
- The conversion of glucose into ATP energy needed by animals presented a fatal problem of spontaneous combustion. If the activation energy necessary to commence the conversion process were supplied by simply adding heat, like chemists do in the laboratory, the living creature would simply burn-up. The development of protein enzymes provided an elegant solution to a fatal problem by increasing the reaction rate of molecules in a chemical reaction without increasing the temperature. That solution, again, overcame a fatal problem presented by the laws of physics.

There is no confirmable data whatsoever in support of the NAS explanations, which contend that these phenomena evolved by the natural selection of **random** genetic mutations. The NAS explanations are inferences based on no scientific evidence. They are explanations spun from hole cloth.

Because these explanations are not based on empirical evidence they are deemed to be doctrine of a religion and cannot be included in science class. They are metaphysical-religious explanations that should be expelled from science class."

Explanation #5

"We are told by Professor Noah Tall that single-celled organisms evolved internal cell structures, with each internal cell structure enclosed within its own protective membrane.

Scientific evidence shows us that these internal cell structures, called organelles, evolved to perform specialized biological functions. Organelles called ribosomes translate the information contained in nucleic acids into the language of amino acids necessary to construct the proteins needed to perform all the functions of life and living. Organelles called lysosomes evolved to rid the cell of the waste products of cellular metabolism. And, an organelle called the nucleus, enclosed within its own membrane, evolved to house and protect the organism's DNA.

Dr. Tall's explanation is that these organelles evolved by endosymbiosis, whereby one single-celled organism was consumed by another and the consumee became a functional part of the consumor.

There is no confirmable data whatsoever in support of that explanation. It is an inference based on no scientific evidence. It is an explanation spun from hole cloth.

Because that explanation is not based on empirical evidence it is deemed to be doctrine of a religion and cannot be included in science class. It is a metaphysical-religious explanation that should be expelled from science class."

Explanations #6 and #7

"Professor Tall explains that all living things contain information within their DNA. The same basic DNA structure is the same for all living things. The basic DNA structure has not evolved and is highly conserved.

He further explains that two other 'highly conserved' features of living creatures originated in the very early years of life on Earth:

- homeostasis,
- the innate immune system.

Plants and animals have the ability to regulate and maintain a stable internal environment within very precise parameters. The ability to maintain such homeostatic conditions is dependent on the ability of the organism to regulate such variables as temperature, blood sugar level, thirst and hunger. The regulation of each such variable within a very narrow range requires the coordinated operation of several independent components:

- stimulus,
- receptor,
- control center,
- effector.

Plants and animals have an innate immune system that allows them to be protected from invading pathogens trying to kill them. This system is dependent on the ability of the cells of the immune system to be able to distinguish both:

- its own body cells from the cells of the pathogen invaders;
- its own normal body cells from its own cells that have been infected by the invader.

Scientific evidence tells us that both homeostasis and the innate immune system developed quite early in the evolution of life. Both homeostasis and the immune system require unique molecules to perform elegant balance procedures and exquisite command and control functions. Yet, Professor Tall explains that both homeostasis and the immune system simply evolved through the natural selection of random genetic mutations that provided a survival advantage to the organism and its offspring.

There is no confirmable data whatsoever in support of the randomness explanation for the evolution of these wonders. It is an inference based on no scientific evidence. It is an explanation spun from hole cloth.

Because that explanation is not based on empirical evidence it is deemed to be doctrine of a religion and cannot be included in science class. It is a metaphysical-religious explanation that should be expelled from science class."

Explanation #8

"We have now reached the point in the evolution of living organisms when multi-cellular organisms developed.

Scientific evidence shows us that the first multi-celled organisms are found in the fossil record after the aggregation of large masses of single-celled organisms called flagellates.

Scientific evidence reveals that shortly before and during the Cambrian Period, 540-510 million years ago, a great leap in evolutionary development occurred. Multi-celled organisms then developed layers of cells, bilateral symmetry, locomotion, a mouth, an anus, a defined body cavity, and a complex nervous system. Again, the fossil record provides scientific evidence for these evolutionary developments. These facts should be taught in science class. However, there is no scientific evidence for how the random mutation of genetic DNA could account for all these developments in a relatively short period, given the snails pace of genetic mistakes.

The scientific explanation provided by Dr. Tall is that all of these developments evolved by natural selection of random mutations of genetic DNA.

There is no confirmable data whatsoever in support of that explanation. It is an inference based on no scientific evidence. It is an explanation spun from hole cloth.

Because that explanation is not based on empirical evidence it is deemed to be doctrine of a religion and cannot be included in science class. It is a metaphysical-religious explanation that should be expelled from science class."

Explanation #9

The Ultimate and Overarching Explanation of Science

"Scientific evidence informs us that the evolutionary developments that we have just reviewed took place over a period of some three billion years. In the roughly 500 million years since the Cambrian Period more and more advanced living organisms have spread throughout the seas and the land masses on the face of the Earth. Scientific evidence from the fossil record, homological comparisons, and DNA analysis tells us that life progressed from plants to insects to fish to amphibians to reptiles to birds to mammals. Along the way none of those categories of living organisms became extinct. And, step by step, as the categories of living organisms got more and more complex, the mechanism that structured complexity was an ever-increasing amount of information content in genetic DNA.

Science has accumulated a wealth of evidence at this time showing that life has evolved, at core, by the increase of information content in DNA. The NAS explanation for all of these evolutionary phenomena is that they developed an increasing amount of information in their genetic DNA. That fact should be taught in science class.

But then the NAS explains that the underlying mechanism for increasing the information content in genetic DNA is **randomness**. The NAS explains that the information in genetic DNA increased as random mutations occurred in the nucleotides of the DNA molecule. They tell us that those mistakes and accidents caused DNA information content to increase, with no intention, direction or purpose involved.

These scientists infer from the fact that the information content of the DNA molecule increased dramatically over the course of the past 500 million years that it had to do so from purely natural causes. Randomness is the only explanation that comports with natural causes. The NAS explains that any explanation other than randomness would not be scientific.

Science has accumulated a lot of evidence concerning the increase of information content in the DNA molecule over the past 500 million years. Mutation rates of genetic DNA have been determined to be very slow. Repair mechanisms within genetic DNA are extremely successful, resulting in an error rate of less than one-in-a-billion. Science has determined that nearly all DNA mutations are either neutral or detrimental in their result.

Laboratory-induced mutations in fruit flies have produced over 3,000 mutants with none of the mutants representing an improvement containing an increase in the information content of DNA.

Further accumulated scientific evidence indicates that the major evolutionary method of increasing the information content of DNA had to be gene and chromosome duplications. Point mutations would be far too slow to account for the increased information content that we observe today. So the discovery of whole gene and whole chromosome duplications was purported to be evidence for the randomness explanation. However, the scientific evidence reveals that the numbers of genetic duplications and chromosome duplications observed are far too few to allow randomness to be a scientific explanation. Another duplication mechanism was needed in order to explain how randomness could have succeeded in producing the results observed. So, the 2R hypothesis was pronounced to be the missing link.

The 2R hypothesis explanation is necessary to explain the timeline that scientific evidence reveals would have been necessary to randomly evolve the information content of DNA that we observe in living organisms today. The 2R-hypothesis explanation is that early-on in the evolution of vertebrates, two rounds (2R) of whole genome duplication **just had to** have happened.

In order to maintain the randomness paradigm, scientists have maintained their ability to provide very creative metaphysical innovations. If a square peg doesn't fit into a round hole, just pound it in with authority.

There is no confirmable data whatsoever in support of the 2R hypothesis. And, there is no confirmable data whatsoever in support of the ultimate and overarching explanation that the change mechanism for the increase of information content in DNA is **randomness**. Those inferences are based on no scientific evidence. Those explanations are spun from hole cloth.

Because the randomness explanation is not based on empirical evidence it is deemed to be doctrine of a religion and cannot be included in science class. It is simply a metaphysical-religious explanation that should be expelled from science class."

* * *

Clarence Darrow returned to the Plaintiff's table, laid his notes upon the table, and turned to again face Judge Raulston before concluding.

* * *

Metaphysical-Religious Explanations Must be Expelled from Science Class

"Your Honor, the ultimate and overarching explanation that the NAS and the School District want Mr. Scopes to teach in science class for each of these amazing biological things is that they occurred naturally through the process of evolution by natural selection of **random** mutations of genetic DNA. They are adamant that science class must include the explanation that everything that lives on Earth is the product of purely natural causes. They insist that science class must teach that life on Earth is ultimately the product of **randomness**, without intention, without direction, and without purpose.

People trust brilliant, gifted scientists to tell them the truth. When our elite scientists proclaim that their atheistic religion is supported by scientific facts they undermine society and science at the same time. Science is necessarily limited to the search for the best naturalistic truth. The mechanism of randomness is not a naturalistic truth. **Randomness** is a **faith-based** explanation. Scientists who explain that the evolutionary mechanism of randomness is a 'scientific truth' are perpetuating a fraud. That fraud supports their atheistic religious belief. But, that fraud is based on absolutely no confirmable scientific evidence.

The foundational belief of elite and mainstream science is that the 'smart information' residing within the DNA of all living things just has to be the product of a simple naturalistic process called random mutation. Yet, the discoveries of science reveal a complexity that cannot begin to be explained by applying the traditional rules of scientific naturalism.

No evidence has ever been adduced by science that random mutation has produced information. Yet, the very foundation of the mainstream version of evolution by natural selection rests firmly on the assumption that not only can unintentional random mutation produce information, but that unintentional random mutation can actually increase information in an exponential manner.

The fraud that is being perpetuated in the science classes of our public schools today in the name of science is the greatest fallacy ever told. That fallacy proclaims that all of the wonders of life **just have to be** the result of no intention and no meaning and no purpose. That fallacy is harmful not only to the students being taught, but also to society at large. Because our most brilliant scientists uniformly teach that fraud as 'true' that fraud has taken on the status of 'The Big Lie'. And, the bigger the lie the more it just has to be 'true'.

With the publication of the books *Science and Creationism* and *Science, Evolution, and Creationism,* the elite scientists of the National Academy of Sciences have delved deeply into metaphysical and religious territory. They clearly espouse a metaphysical belief of expanded naturalism, called scientism, whereby the only possible truth must derive from natural causes. And, thereby, they clearly espouse a religious belief called 'scientific atheism'.

When our most brilliant scientists proclaim that scientific discoveries 'prove' the idea that life began and evolved without purpose or intention or direction they send to us the clear message that life itself is without purpose and devoid of meaning.

Every reputable scientist in America aspires to be elected to the membership of the National Academy of Sciences. The NAS is the most highly esteemed scientific organization in this country. Sadly, the bias of the NAS membership against traditional religion has resulted in the elite scientists of the NAS becoming the prophets of 'scientific atheism'.

In America everyone is entitled to his or her own religious belief. That includes atheistic belief. The true danger for society presented in 'scientific atheism' is the inclusion of the word 'science'. The overwhelming implication is that it presents an authoritative 'truth' that derives from scientific discoveries. The randomness doctrine of 'scientific atheism' is now included in science textbooks and taught in the science classes of public schools. No other religious belief is afforded that opportunity, and rightly so. And, 'scientific atheism' should not be afforded that opportunity either, but it is.

Doing so places the religion of 'scientific atheism' in a favored class all its own. It provides 'scientific atheism' with the imprimatur of discovered 'truth'. And that is simply wrong.

The insidious effect of 'scientific atheism' is to devoid the human psyche of the hope and inspiration of higher purpose and meaning in life itself.

Allowing the randomness doctrine of 'scientific atheism' to be taught in our public schools is nothing more complicated than the state allowing one favored religion – 'scientific atheism' – to be placed above all others. That is simply wrong. And that is simply in violation of the United States Constitution.

Thank you."

* * *

Judge Raulston allowed Clarence Darrow to gather his papers together and return to his seat before he asked William Jennings Bryan to proceed with his closing argument. The seasoned Defense counsel then rose from his seat and began.

* * *

The Defendant's Closing Argument

"Thank you, your Honor.

It is really most difficult for me to mount an argument on the merits against the spurious accusations made by Mr. Darrow against Professor Tall in particular and the National Academy of Sciences in general. That is because Mr. Darrow's arguments are completely without merit. The very idea that these distinguished scientists are advocating the preaching of a religion in public schools is ludicrous and scandalous.

The National Academy of Sciences includes about 1,800 members who provide a valuable public service to this country. These distinguished men and women follow the tradition of public service begun under the administration of President Abraham Lincoln when he signed the National Academy of Sciences Act of Incorporation in 1863. The members of the NAS are elected by their scientific peers in recognition of the outstanding achievements that they have attained in their specialized fields of science. The integrity of the distinguished scientists of the National Academy is beyond reproach. To assert that they are furthering some sort of atheistic religious agenda is preposterous at best.

Dr. Noah Tall is the President of the NAS. He has been elected by the membership to lead the Academy because of his accomplishments in the field of microbiology, his dedication to the expansion of scientific knowledge, and his leadership skills. Professor Tall earned his bachelors degree in biology from Harvard University, his M.D. from Johns Hopkins, and his Ph.D. in Molecular Biochemistry from Oxford University. His research has produced remarkable results in the prevention and cure of numerous diseases and medical conditions. Indeed, he has been recognized for his work by many awards, including the Nobel Prize in Medicine. Mr. Darrow's characterization of Professor Tall as a religious zealot who is championing the teaching of some sort of atheistic religion in public schools is both without any substance and highly offensive."

* * *

William Jennings Bryan had delivered the preceding while at the lectern. He then began to pace and stop and pace and stop in front of both the Plaintiff's table, the Defendant's table, and in front of the empty jury box, as he continued with the following.

* * *

"Science is process. It is a painstaking step-by-step process. Many of the intricacies of scientific knowledge are dependent on an understanding of higher mathematics that most people simply do not understand. Most people simply do not have the education in higher mathematics necessary for understanding. Many of the intricacies of scientific knowledge are dependent on an understanding of physical causal relationships of physics and chemistry and biology that most people simply do not have the education necessary for understanding. Because of this fact the courts of this country have a long history of reliance on the testimony of witnesses who are tested and qualified to testify as expert witnesses in their particular field.

Dedicated scientists devote their very lives to the study of mathematics and physical causal relationships in order to gain insight and understanding necessary to qualify them as expert in their field. They then use that insight and understanding to better our everyday lives. These dedicated men and women deserve our gratitude, not our derision.

Mr. Darrow is a layman. I am a layman. Time and time again in his questioning of Professor Tall, Mr. Darrow has derided the Professor's answers. His derision is simply the result of his failure to understand the special insights into the reality of things physical that Professor Tall has gained through his many years of dedicated scientific service. It is not Dr. Tall's shortcoming that prevents Mr. Darrow from understanding complex scientific phenomena without having the necessary background in the scientific method and mathematics. The shortcoming is Mr. Darrow's. Mr. Darrow is not the expert witness in this trial. Dr. Noah Tall is.

Expert witnesses are recognized in our legal system as invaluable assets to finding the judicial truth in the courtroom. There is no higher scientific authority in America than the National Academy of Sciences. And, it has been stipulated that Dr. Noah Tall, as the President of the NAS, is the most highly qualified expert witness possible to provide expert scientific testimony in this trial.

In a murder trial an expert medical witness is needed to determine the cause of death. In the same trial, an expert in the science of DNA analysis is needed to determine if the blood on the knife is that of both the victim and the accused. An expert in the science of fingerprint analysis is needed to determine if the fingerprints found on the knife are those of the accused. These experts provide testimony regarding things that laymen could not possibly figure out by themselves. This expert testimony is essential to furthering the interests of justice.

In this trial Dr. Noah Tall has been providing expert witness testimony on a wide range of scientific subjects that laymen could not possibly figure out by themselves. Just as scientific expert witnesses are needed to testify to the accuracy of DNA and fingerprint evidence in a murder trial, Dr. Tall has been needed in this trial to testify to the accuracy of scientific discoveries regarding the mechanism of evolution by natural selection."

* * *

William Jennings Bryan positioned himself at the front edge of the Defendant's table so that he could gesture repeatedly toward Clarence Darrow as he proceeded.

* * *

"Mr. Darrow has contended that the NAS belief in the ultimate reality of randomness is not based on scientific evidence. He says that such a belief is metaphysical. That contention is simply bunk. There is an overwhelming amount of evidence for randomness, as Professor Tall has repeatedly testified. Professor Tall and the NAS are not concerned a wit with ultimate answers and ultimate reality, the stuff of metaphysics.

Mr. Darrow says that the two NAS books specified by the Dayton School District for use in science class call for teaching atheism in public school. That is simply bunk. The NAS wants only to teach the evidence produced by science. Nobody wants to preach atheism. They simply want to teach the facts. The mechanism of randomness in nature is a scientific fact. Students can draw their own conclusions from that scientific fact.

Mr. Darrow says that the NAS preaches the 'doctrine of randomness' based on no evidence. That is simply bunk. Nothing could be further from the truth. Time and again Dr. Tall has provided expert witness testimony that, in and of itself, supplies overwhelming evidence for randomness in nature.

Mr. Darrow points out that photosynthesis is an elegant and complicated process. So what? Professor Tall has testified as an expert witness that elegant and complicated organic processes evolve by the natural selection of random genetic mutations. Who is the expert here, Mr. Darrow or Professor Tall?

Mr. Darrow points out that maintaining electrical charge in animal organisms is an elegant and complicated process. So what? Dr. Tall has testified as an expert witness that elegant and complicated organic processes evolve by the natural selection of random genetic mutations. Who is the expert here, Mr. Darrow or Dr. Tall?

Mr. Darrow says that there is no evidence behind the 2R hypothesis that scientists use to support the evolution of more complicated animals through the process of gene and chromosome duplication by random mutations. So what? Just because there is no absolute proof for the hypothesis yet does not mean that it is not a perfectly reasonable scientific explanation for the physical realities that have been observed by science.

Mr. Darrow nit picks that scientists have, as of yet, not been able to create life from non-life in the laboratory. So what? Scientists have discovered a great deal about how nonbiological components show great promise for the explanation of how life first began on Earth. Although all the answers are yet to be found great strides have already been made. Science is a painstaking step-by-step process.

The step-by-step discovery of evidence based on facts about the natural world is the path of science. Mr. Darrow's contention that the brilliant scientists of the NAS are the prophets of scientific atheism is an appalling statement based on no evidence whatsoever.

The mechanism of randomness is not some sort of religious doctrine. Randomness is a proven scientific truth as attested to by expert scientific testimony. The expert testimony of Dr. Noah Tall has provided all of the evidence that any reasonable person needs to support that scientific truth.

Your Honor, on behalf of Professor Noah Tall and the distinguished scientists of the National Academy of Sciences, I ask that you provide a swift decision in this case. Summarily put an end to this farcical lawsuit and reaffirm the position of authority and stature that has been earned by the dedicated scientists of this country. Thank you."

* * *

Judge Raulston allowed William Jennings Bryan to regain his seat at the Defense table. He then addressed the observers in the courtroom.

"For the benefit of those in the audience who are not accustomed to following trials in Federal Court let me explain before continuing.

In a Federal civil trial both the counsel for the Plaintiff and the counsel for the Defendant are provided an opportunity to make a closing argument. But, lastly, the Plaintiff's attorney is afforded the opportunity to make a final statement. The purpose of that is to allow the party who brought the lawsuit to have the final word.

Mr. Darrow, you may proceed with your final statement."

* * *

Clarence Darrow took his place before the lectern and began.

The Plaintiff's Final Statement

"From the beginning of recorded history until the adoption of the Bill of Rights contained in the Constitution of the United States of America, civil society for each tribe or City-state or nation was inexorably intertwined with a fundamentalist religious belief. Membership in civil society required one to believe in the ultimate reality – God – prescribed by the State. Government and religion were one.

The genius of the United States Constitution, with the adoption of the Bill of Rights in 1791, was to break the bonds between government and religion. The simple words of the First Amendment to the Constitution provided that, in this country, for the first time in the history of the world, the civil government would never involve itself in the religious affairs and beliefs of the people:

> 'Congress shall make no law respecting an establishment of religion, or prohibiting the free exercise thereof . . . . '

From the adoption of that First Amendment unto this very day, the Supreme Court of this nation has rigorously enforced the novel idea contained in those few simple words. But for one glaring exception. That exception is the centered heart of this lawsuit. And, that exception provides historical throwback to again reunite government with a fundamentalist religion.

The fundamentalist religion that is now tacitly embraced by this nation's government is not a traditional Islamic or Judaic or Christian religion. No, the State-sponsored fundamentalist religion that is now tacitly embraced by the Law Courts of this country is the religion of scientific atheism.

The fundamentalist religion of scientific atheism is solidly founded on an unyielding belief in the power of accident and randomness. The religious doctrine of randomness is a doctrine of this religion that simply **must be true** if one is to dutifully believe in the ultimate reality of the Accidental God of scientific atheism – Darwin's God."

* * *

Clarence Darrow returned to the Plaintiff's table to obtain some additional notes, returned to the lectern, and continued.

* * *

"Let's return our focus to the Tennessee statute entitled the *'Only Science in Science Class Act'*. It specifies:

'The teaching of science in science classes shall be limited to only natural explanations that can be inferred from confirmable data whose results can be observed, tested, replicated and verified. Scientific explanations are restricted to those that can be inferred from confirmable data. Inferences that cannot be inferred from confirmable data are deemed to be doctrine of a religion and cannot be included in the science curriculum of public schools. Explanations that cannot be based on empirical evidence are not a part of science.'

The explanation of **randomness** is extensively used by the NAS in their two books entitled *Science and Creationism* and *Science, Evolution, and Creationism*. Those are the books that the Dayton School District has mandated must be a part of the biology curriculum. In these books the NAS explains that **random** genetic mutation is the mechanism whereby information content is increased in DNA that allows for the evolution of more and more complicated living organisms.

Science has discovered overwhelming evidence that mutation of genetic DNA is the mechanism whereby the information content of DNA is increased that sustains evolution. That is not in dispute.

But, has science actually discovered any evidence at all to sustain the belief that such DNA mutations are **random**?

This then is the pointed question. Does the explanation of **randomness** qualify as a scientific explanation that can be taught as true in science class?

The evidence produced in this trial has clearly revealed that the answer to that question is no. There is not a scintilla of scientific evidence in support of the randomness explanation that our most elite scientists insist is the mechanism by which life began and through which adaptive life has evolved. The randomness explanation is not a scientific explanation. It is a metaphysical-religious explanation that has no place in science class.

The metaphysical-religious explanations of our traditional religions would inform us that the mechanism that sustains the evolution of living organisms by increasing the information content of DNA is the work of a Purposeful God. The metaphysical-religious explanations of our most elite scientists insist that the mechanism of randomness is the work of the Accidental God – Darwin's God. Neither of these metaphysical-religious explanations should be taught as true in science class."

* * *

Clarence Darrow paused to take a few sips of water and to organize his notes before continuing.

* * *

"Science has discovered that DNA information is contained in a coded language. And, science has discovered that mutations have provided for an increase in the coded information content residing within DNA molecules. Those changes in DNA resulted from the mechanism of either randomness or purposefulness.

Logic and commonsense would allow us the inference that it is not reasonable for the mechanism of randomness to first develop a complicated coded language and then to increase information content through randomness.

But, logic and common sense are not the stuff of science. The scientific method and the Tennessee statute demand that an explanatory inference must be based on confirmable empirical data in order for the explanation to be considered scientific.

There is absolutely no confirmable empirical data to support either the inference of purposefulness or the inference of randomness as the mechanism that developed a coded DNA language and then was used to increase DNA's information content. Neither inference should have a place in science class."

* * *

The Plaintiff's attorney began to walk, with the aid of his cane, back and forth before the Judge's bench.

* * *

"The foundational explanation that is today taught in the science classes of our public schools is that each and every organ and each and every metabolic system within each and every living organism **just had to** be evolved by the natural selection of **random** mutations of genetic DNA.

Our eyes and ears and hearts and kidneys and livers and brains and blood circulation and clotting systems and our intricate immune system **just had to** develop through the natural selection of mistakes and accidents.

When the DNA of a virus or bacteria mutates rapidly to develop resistance to antibiotics that our brilliant scientists have developed to kill those attacking pathogens, we are told that they do so by a random process. Logic and common sense tells us that a single-celled organism cannot accidentally outwit our brilliant scientists. But, those same brilliant scientists tell us that, indeed, those pathogens simply solve the problem facing them by accident, by random mutation of genetic DNA. We are told that not on the basis of scientific evidence, for there is no such scientific evidence. We are told that because **randomness** is the explanation that comports with the scientific method.

The **randomness** explanation may comport with the scientific method. But that does not make randomness a scientific explanation. A scientific explanation must be testable and verifiable. **Using the mechanism of randomness to explain both the conception and evolution of all life on this planet is neither testable nor verifiable. That explanation has no place in science class."**

* * *

Clarence Darrow returned to the lectern. After a pause of almost a minute to consult his notes, he continued.

* * *

"In the final analysis, scientific inferences must be reasonable. Reasonable is defined as moderate and fair; agreeable to reason; not extreme or excessive.

Nothing about the mechanism of randomness provides a reasonable inference. The NAS has used the mechanism of randomness to explain things without evidence. That is not moderate and fair. That is not agreeable to reason. That is extreme and excessive. That is the stuff of metaphysics and religion. That is not the stuff of science.

Very few lawyers and very few judges in America are scientists. And, the underlying problem for us non-scientists is that this scientific atheistic belief is clothed in the garment of science. We tend to believe scientific authority. We believe these scientific authorities based not on the evidence that these eminent scientists provide. Rather, we accept their pronouncements based on their authority alone. We trust elite scientists to tell us the truth.

As Professor Tall has stated repeatedly, the only scientific explanation **just has to be** a natural one. And the only possible natural explanation for how the information in DNA increases as life evolves **just has to be** random mutation. If the mutation, the change that produces an increase of information content in DNA, were found not to be random, the only other possible explanation would be non-random. Non-random, of course, means that the information was, in some unknown way, intentional, directed, and purposeful. For these elite scientists, that explanation simply cannot be true. Indeed, these dedicated and brilliant scientists are **absolutely certain** that no intention or direction or purpose was involved in the creation of and in the evolution of life on Earth. Such absolute certainty is the stuff of religion, not science.

These brilliant scientists are certainly entitled to their belief. This is America, and in America each of us, scientist and layman alike, has the inalienable right to hold any metaphysical and religious belief that we choose to. But, metaphysical and religious beliefs should not be taught in the science classes of our public schools. That practice is in violation of our Constitution."

When Science Becomes Religion

"Learned counsel for the Defendant has strenuously objected to the very idea that the brilliant scientists of the NAS want to preach a religion in public schools. He says that such is both ludicrous and scandalous, that what these scientists teach is the opposite of religion. Such a characterization is in fact a reference to nonreligion. And the United States Supreme Court in *Epperson v. Arkansas* held that:

> 'The First Amendment mandates governmental neutrality between religion and religion, and between religion and nonreligion.'

In reality, such a nonreligion characterization is just another name for 'scientific atheism'. A rose by any other name is still a rose. Belief in the ultimate reality of randomness is a core doctrine of the religion of 'scientific atheism'.

Darwin's theory provided scientists with the crude ammunition to mount a full-scale scientific revolution that would result in changing a worldview of created order into a worldview of happenstance. The mainstream scientific community thus adopted an *a priori* random evolutionary bias for everything. **Randomness** became a scientific requirement for everything. Scientists proceeded to establish a credo of impregnable axioms and tenets in support of the new dogma that has now become firmly established in the religion of scientific atheism. Doctrine of the religion of scientific atheism proclaims that there is no need for a creative intelligent agent of the universe. Homage is dutifully given to the **Accidental God – Darwin's God.**

Scientific atheism redirected the goal of science. Historically, the goal of science has been to search for the most truthful explanation based on evidence from the physical world. But now, the search is restricted to the best **random** naturalistic explanation. If design appears everywhere in the natural world it is of no consequence or significance. No amount of apparent design can support a conclusion of actual design. Actual design is forbidden by the doctrine of randomness, which is a core belief of the religion of scientific atheism."

Government and Religion Must Again be Separated

"The religion of scientific atheism developed and grew as an unplanned and unintended consequence of scientific devotion to the scientific method. The 'old' worldview of not-by-chance and abundant purpose was abandoned by the elite scientific community. It was replaced by the 'modern' worldview of by-chance-alone and no purpose. The 'modern' worldview now holds that the natural world and everything therein evolved purely by chance without any design or purpose. The new religious dogma required that **randomness** underlie all scientific explanations. And, that is precisely the religious doctrine that is not just allowed, but that is now required to be taught today in public school science class.

This religious doctrine, which is now taught as true in our public school science classes, informs our children that there is no special significance to our lives at all. We are simply the product of chance and circumstance. As such, there is, at core, no real purpose and meaning to the universe or anything therein, including us. There is no purpose and meaning in life at all except for that which we simply make up. Our existence, in essence, is simply the result of some cosmic hiccup. That is what the religious doctrine of the elite scientific community says. And, that religious doctrine, supported by so many brilliant scientists, has a profound effect on the rest of us."

* * *

Attorney Darrow retrieved some additional notes from the Plaintiff's table and returned to the lectern. He continued.

* * *

"Scientific advancements have been made by the rigid adherence to the scientific method underpinned by an insatiable quest for truth. That is the nature of good science.

Our most brilliant scientists have made incredible discoveries about the physical world. They have thereby developed many wondrous applications and inventions that provide for our great comforts of modern life. Their work allows us to live longer and to prosper far better than our forebears.

But, there is much more to be discovered. And, much more that may remain forever undiscovered. It may well be that some of the fundamental intricate laws of the universe and life are so intricate that mere human intellect may forever be incapable of discovering them. There is certainly no logical reason to assume that we humans, ourselves the products of stardust, should be able to factually discover the ultimate realities of the universe or of life.

Brilliant scientists serve as our knowledge authority figures for things physical. We trust that they tell us truthfully only the results of good science. When they delve into the world of metaphysics and religion they owe a duty to the rest of us to tell us just that.

Today the group-think of atheistic scientists has developed a paradigm of scientific belief that is akin to the group-think of theistic scientists past. At one point or another in human history the elite and mainstream majority of scientists believed in such things as the 'truth' of:

- all outer space beyond our atmosphere being composed of a mysterious 'ether';
- an Earth-centered solar system;
- 'pangenes' of a mother and father 'preforming' a tiny, fully-formed person that simply grew larger within the mother's womb; and
- spontaneous generation of life from non-living matter, whereby frogs grew out of mud, rats grew out of garbage and flies were born from rotting meat.

In our past history the elite and mainstream majority of scientists believed in these things and maintained that they were true. They were absolutely certain that they were right. And, they were dead wrong. Yet they belittled and chastised anyone who challenged their certitude.

The group-think of the elite and mainstream majority of scientists past caused them to actually see a smaller epicycle within a larger epicycle in order to maintain belief in an Earth-centered solar system. The group-think of the elite and mainstream majority of scientists past caused them to deny the wonders of embryological development and actually see a tiny preformed person residing within an egg or sperm sex cell and to actually see rotting meat giving birth to flies.

The group-think of the elite and mainstream majority of scientists today causes them to actually see mistake compiled upon mistake, again and again and again, resulting in the development of exquisite living systems that are capable of intricate communications and command and control functions. They are absolutely certain that they are right. Randomness **just has to be** true. And, if a square peg doesn't fit into a round hole just pound it in with authority.

Absolute certainty is the stuff of fundamentalism. And, fundamentalism is the stuff of religion, not of science.

Yet, as of this date, our Courts have not recognized the religious nature of the doctrine of randomness now taught as true in our public schools. The failure to do so has the insidious effect of tacitly establishing a new State-sponsored religion. The establishment of such a State-sponsored religion is clearly in violation of one of the most fundamental precepts of our Constitutional liberties.

The Randomness Doctrine Establishes a
State-Sponsored Religion

In our society a Court of Law is the objective forum that is charged with the task of resolving disputes. Judges and juries listen to the evidence produced by both parties to a dispute and then decide which side is better supported by the evidence produced.

There is today in this country a real dispute among scientists over whether the creation and adaptive evolution of life on this planet is the product of a purposeful and directed process, dubbed Intelligent Design (ID), or a simple process of randomness acting without direction or purpose.

The vast majority of elite and mainstream scientists favors the theory of evolution by natural selection of **random** mutations of genetic DNA. A much smaller minority of scientists favors the theory of evolution by natural selection of **purposeful and directed** mutations of genetic DNA. This minority group has no idea whatsoever of what the purposeful and directive force that drives evolution by natural selection is. But, they are convinced that there is no way that all of the wonders of the living world could have evolved by nature selecting from a purely random array of mutations.

To date, our Courts have examined this dispute in several different formats. They have always determined that the religious Establishment Clause of the Constitution bans any purposeful and directed answer to the dispute. This ban has resulted in a *de facto* default judgment that has placed the randomness explanation beyond the realm of religion.

That conclusion made a lot of sense when the Court relied solely on expert scientific **opinions** as the evidential measurement yardstick. The preponderance of expert testimony has clearly been found on the side of randomness. When expert numbers and credentials are added up as the sole measure of evidence, the randomness explanation is clearly the scientific winner. The Courts have certainly made no error in that regard. The randomness explanation is supported by the vast majority of our elite and mainstream scientists.

Yet, a huge error of reasoning has resulted that directly affects Constitutional rights. The error has been in the evidential measurement yardstick used. And, that again has been no fault of the Courts that have examined the issue to date.

A Court of Law rightfully limits a dispute to the pleadings filed by the parties to the dispute. The Court does not expand the dispute beyond what the parties plead. The Court then resolves such a limited dispute based solely on the evidence presented by the parties to the dispute.

To date, each dispute regarding the violation of the religious Establishment Clause has been plead in a fashion that limits the dispute to whether **traditional** religious explanations can be taught in public schools. Courts have banned the Biblical creation story, the teaching of creation science and the theory of Intelligent Design (ID) from the public school science classroom. Those banishments have been most appropriate as violations of the religious Establishment Clause. However, the default understanding that has been produced by those banishments has been a tacit approval of teaching randomness as true in science class.

The instant lawsuit before this Court today challenges that default understanding and tacit approval for the first time. And, the instant lawsuit challenges the expert witness testimony itself as a true evidential measurement yardstick.

The lawsuit before this Court today demands that the actual measure of evidence can no longer be based on the **opinion** of scientific experts that is based only on their metaphysical and religious beliefs. The actual measure of evidence must be **confirmable empirical evidence** that is actually presented before this Court. That is the very standard prescribed by Tennessee's *'Only Science in Science Class Act'*, and that is the standard that must be used by this Court.

To date the randomness explanation, supported by the vast majority of our elite and mainstream scientists, has been allowed to be taught in our public schools as true. That **randomness belief** is every bit as much a metaphysical and religious belief as the **purposeful belief** of traditional religions. That fact has never before been challenged in a Courtroom before this lawsuit. Now that it has been challenged, the randomness explanation must be expelled from the public school science classroom along with the expulsion of all other religious doctrines.

The failure to expel the doctrine of randomness from science class inexorably entwines government and religion. That may have been nobody's intention. But that is the clear result. And, that result is anathema to the principles of the freedom of religion and the separation of Church and State upon which our nation was founded.

Government and religion must again be separated by expelling Darwin's God from the science classrooms of our public schools."

* * *

Clarence Darrow paused for several moments and took a drink of water. With the assistance of his cane he walked in front of the Plaintiff's table and finished his final statement in the trial, directly addressing Judge Raulston.

* * *

Both Intelligent Design and Randomness
are Religious Explanations

"In 1968 the Supreme Court of the United States, in the *Epperson* case, made it crystal clear that metaphysical-religious doctrines have no place in our public schools:

> 'Government in our democracy, state and national, must be neutral in matters of religious theory, doctrine, and practice. It may not be hostile to any religion or to the advocacy of no-religion; and it may not aid, foster, or promote one religion or religious theory against another or even against the militant opposite. The First Amendment mandates governmental neutrality between religion and religion, and between religion and nonreligion . . . .
>
> The State may not adopt programs or practices in its public schools or colleges which "aid or oppose" any religion. This prohibition is absolute. It forbids alike the preference of a religious doctrine or the prohibition of theory which is deemed antagonistic to a particular dogma.'

In 1987 the Supreme Court of the United States, by the ruling in *Edwards v. Aguillard,* made it crystal clear that:

> 'The First Amendment does not permit the State to require that teaching and learning must be tailored to the principles or prohibitions of any religious sect or dogma.'

The evidence produced in this trial has clearly shown that the *'Only Science In Science Class Policy'* of the Dayton School District provides an undeniable preference for the religious doctrine of randomness. And that the policy of the school district requires that teaching and learning must be tailored to the principles and prohibitions of the religious sect of scientific atheism.

In 2005 Judge Jones of the United States District Court for the Middle District of Pennsylvania ruled in the case of *Kitzmiller v.*

Dover Area School District that the mechanism of Intelligent Design (ID) could not be taught in science class. He found that:

> 'ID fails to meet the essential ground rules that limit science to testable, natural explanations.'

Today we ask you to rule that the mechanism of **random** mutations of genetic DNA also cannot be taught in science class. The explanation of **randomness** likewise fails to meet the essential ground rules that limit science to testable, natural explanations. The mechanism of randomness should not be taught in public schools. Both purposefulness and randomness are metaphysical-religious explanations that have no place in science class.

In the *Kitzmiller* case Judge Jones concluded that:

> 'It is our view that a reasonable, objective observer would, after reviewing both the voluminous record in this case, and our narrative, reach the inescapable conclusion that ID is an interesting theological argument, but that it is not science.'

Your Honor, we ask that after reviewing the record in this case, you likewise conclude that a reasonable, objective observer would reach the inescapable conclusion that **the explanation of randomness is an interesting theological argument, but that it is not science**.

The *'Only Science in Science Class Policy'* of the Dayton School District requires Mr. Scopes to include in his science class the **metaphysical-religious explanation of randomness** that is prophesied by the National Academy of Sciences. We ask that you determine that Policy to be unconstitutional and void as a violation of the First Amendment of the United States Constitution.

Thank you Your Honor."

EPILOGUE

After closing arguments and Clarence Darrow's final statement Judge Raulston adjourned the proceedings. As both the trier of fact and ruler of law the Judge would now sift through all the evidence, write a legal opinion and issue a ruling in the case.

* * *

I wrote this little book after Judge Raulston took the case under advisement and before he issued his legal opinion and ruling. That's my story and I'm sticking to it.

As a matter of fact, I would not presume to know what the actual ruling in this case would be. I am not skilled enough in the law or in opinion writing to present Judge Raulston's ruling.

Putting legal matters aside, it is not really all that difficult to see that ultimate reality – God – is either purposeful or accidental. In the final analysis each of us worships either a Purposeful God or an Accidental God, also known as Darwin's God. And, these metaphysical explanations for the creation and evolution of life represent polar opposite worldviews.

When the actual discoveries of science are examined in more and more depth it becomes clearer and clearer that the actual mechanism that changes the nucleotide sequences of DNA to provide more and more information content is either purposefulness or randomness. Neither change mechanism can be inferred from confirmable empirical data. And, explanations that are not based on confirmable empirical data should not be taught as true in science class.

The National Academy of Sciences firmly believes that the explanation of randomness is a scientific fact, that randomness is a discovered scientific 'truth'. And, the NAS wields enormous scientific authority in this country.

It is most difficult to believe that Judge Raulston, or the Supreme Court of the United States for that matter, would ever be able to conclude that the atheistic religion of our scientific elite, resting solidly on the foundation of randomness, cannot be taught in science class.

Maybe some day we will see. In the meantime, it is back to Hazel for me. I've got some gardening tips to write for Wednesday's paper.

Best Regards,

Shorty Story